The Way of a Ship in the Midst of the Sea

The Life and Work of William Froude

by

David K. Brown RCNC

There be three things too wonderful for me, yea, four which I know not,

The way of an eagle in the air;
The way of a serpent upon a rock;
The way of a ship in the midst of the sea;
And the way of a man with a maid.

Proverbs 38–18, 19

© D. K. Brown 2005

First published in 2006 by
Periscope Publishing Ltd
33 Barwis Terrace
Penzance
Cornwall TR18 2AW

Copyright © D. K. Brown

All rights reserved. No part of this book may be reproduced or transmitted in any form or by any means, electronic or mechanical, including photocopying, recording, or by any information storage and retrieval system, without permission in writing from the publisher.

A CIP record for this book is available from the British Library

ISBN No 1-904381-40-5

Printed and bound by CPI Antony Rowe, Eastbourne

Frontispiece – The Admiralty Experiment Works 100 years on.
Edmund's tank is left bottom. Later long tank is at rightangles in the foreground. Behind is the manoeuvring tank (400 x 200).

The Way of a Ship in the Midst of the Sea

Contents

Foreword i

Introduction ii

A Note on Sources v

Acknowledgements x

Chapter 1 – William Froude – Family and Early Life 1

Chapter 2 – Railway Engineer 17

Chapter 3 – Early Retirement 27

Chapter 4 – Rolling 1857–1870 39

Chapter 5 – Rolling 1870 – 1875 55

Chapter 6 – A Sacred Duty to Doubt 71

Chapter 7 – Paignton, Chelston Cross and the Brunels 87

Chapter 8 – Ship Hydrodynamics, Background 105

Chapter 9 – SWAN and RAVEN Experiments, 1865–1867 117

Chapter 10 – Admiralty Approval 127

Chapter 11 – Mr. Froude's Tank at Torquay 143

Chapter 12 – The Greyhound Trial 1872 169

Chapter 13 – Years of Achievement 1872–79 177

Chapter 14 – Epilogue – Edmund Froude 1879–1919 215

Chapter 15 – William Froude – An Evaluation 239

Index 251

The Way of a Ship in the Midst of the Sea

Foreword

I was first introduced to the works of William Froude and his son, Edmund, as an undergraduate in 1948. As then presented, their work seemed dull, leading to a mechanistic procedure for estimating the hydrodynamic performance of ships. In my first job on warship design I discovered that experts could differ widely in their interpretation of model tests and my increasing interest in the subject was rewarded with an appointment to the Admiralty Experiment Works, Haslar, (AEW) the research establishment founded by the Froudes (now part of Qinetic).

Working mainly on propeller design, I soon discovered that the Froude tradition was alive and well and involved the application of the most advanced theory, model testing, full scale trials* and a touch of intuition to the solution of current problems. Lecturing on ship hydrodynamics at the RN College and later at University College, London, reinforced these views. A return to Haslar as Chief Constructor coincided with the Centenary Open Days, for which I was responsible, with many displays of the establishment's most advanced work. There was also a historical exhibition that proved so popular that it was developed into the Froudes' Museum§.

More recently, serious study of 19th century warship design led to a growing realisation of the importance of the work of the Froudes and, on retirement, I was able to further my studies of their work with the aid and encouragement of Professor Buchanan and Bath University who granted me a Rolt Fellowship.

William Froude's other interests: railway engineer, the design of agricultural machinery and, above all, his correspondence with John Henry Newman (later Cardinal) form part of a pattern showing how his mind worked. Readers should be aware that the author is at least as agnostic as Froude became which, despite care, may lead to some bias in treating this correspondence. He was a very lovable man, admired by Admirals and Dockyard mateys alike, whilst his wide range of friends, particularly Isambard and Henry Brunel, contributed much to his achievement. I have tried to concentrate on how he worked and on the continuing value of his work with a very few more technical passages, mainly as end notes. The details of his methods may be found in his technical papers.

---

* There is a useful tradition at AEW, which I have followed that tests refer to models, trials to ships

§ Since this book was written, the Froudes' museum has been closed. The artefacts and documents are held by the Science Museum in store.

Introduction

Writing at the beginning of the new century, on naval developments of the 19th century, Nathaniel Barnaby, Director of Naval Construction (DNC), 1870–1875, said of Froude

> '... to him the nineteenth century owes a very large part of its distinction. If the author had to choose a man whose portrait should be in the front of these pages as more representative than any other of the foremost workers in Naval Development during a wonderful century he thinks he would find himself supported by the shipbuilders all over the world in choosing Mr Froude.'

Barnaby was in charge of warship design for the Royal Navy at the time when Froude was just beginning to produce useful results and the passage quoted above is written with a full understanding of the value of William's work. William White, Barnaby's successor, was to make similar comments, quoted in Chapter 14, on the value of Edmund's work.

In the early days of the steamship there was no way of estimating the power required to drive the ship at a given speed and there were many cases in which the speed achieved was disappointing or in which the ship was burdened with engines which were unnecessarily powerful, heavy and hence expensive, also reducing the payload which could be carried. It was also believed that there was a single ideal form for all ships at all speeds. Rolling was a fact of life at sea but not understood and about which it was believed that nothing could be done. Froude's investigations illuminated both problems and provided practical means of achieving the desired performance. Within a very few years reliable estimates of speed and power could be made, a theory of rolling in waves was established and practical measures to limit roll were established.

The Froudes published all their theoretical work and procedures and directly assisted many other countries and companies to set up their own research establishments so that today there are some 150 ship model test tanks, world wide, using methods derived from William Froude's pioneer work and the great majority display a portrait of Froude in the entrance hall or director's office. Several new tanks have been christened with a flask of water from Froude's tank.

Over the last century, vast quantities of fuel have been saved by the application of model testing. Even today a 5% improvement is likely following tests – and this is an improvement over a form which is already quite good since it will be derived from the tabulated results of earlier tests. This may not sound much but in 1980 the Royal Navy alone used 833,000 tons of oil at a price of £150–200 per tonne – 5% of £150 million is not negligible and yet it is a small part of the value of the work to the Royal Navy let alone that to other navies and

Introduction

commercial operators. The work on rolling has been extended to other aspects of ship motion and this has probably led to much greater savings in terms of avoiding loss of operational capability.

It is very greatly to the credit of the Admiralty and, in particular, to Edward Reed, Chief Constructor of the Navy, that they recognised the value of Froude's work and contributed a comparatively large sum to support his investigations. The Admiralty Experiment Works, as Froude's tank became known, was not quite the first Admiralty research establishment being ante dated by the Royal Observatory and the Admiralty Chemist's Department.

Froude is important in another sense. Much of his correspondence with Isambard and Henry Brunel and with J. H. Newman has survived and these letters give an unusually complete picture of the way in which he worked. William Froude frequently quoted the phrase 'Probability is the guide to life' and became increasingly agnostic, referring to the 'Sacred duty to doubt'. In consequence, his prolonged debate with Newman on the nature of proof and on scientific method are particularly relevant.

Froude's life falls into three main periods. In his early life he showed outstanding academic ability, first at school and then at Oxford, whilst his interest in sailing was already marked. After graduating, he worked as assistant to two great railway engineers, first Palmer and then I. K. Brunel. Increasingly, he won the respect of Brunel, first as an effective manager and later, for his diplomacy in soothing local opposition to new lines. He made a number of small but significant advances in widely different aspects of railway technology, often over throwing accepted rules.

Then, aged 36, he gave up regular professional work for a decade in what is usually the most productive era of an engineer's career. Allegedly, he was looking after his ailing father, but in a well-off household with plenty of servants this can hardly have been an arduous task and his father's long life suggests that his illness was not acute. William Froude was very active during this decade as a magistrate, trustee and particularly in judging machinery at agricultural shows while he also carried out some preliminary experiments on ship propulsion.

In 1857, I. K. Brunel tempted him back to work and within four years he had changed the whole approach to the behaviour of ships in waves. He then turned to the problem of estimating the power needed for a new design and, despite the lack of success by previous experimenters, he believed that model tests could be used to give reliable results. In the mid 1860s Froude carried out some tests which convinced him and Reed that model results could be applied with confidence and the

Admiralty paid for the building of a test tank. There was then a short period before his death in which he literally changed the shape of the Navy.

Froude acknowledged the influence of three great men on his life; his older brother, Hurrell, I. K. Brunel and J. H. Newman seeing them as having much in common but also as complementary. He was also influenced by his mother and his wife. The former, Margaret Spedding, was widely praised for imagination, intellect and beauty whilst his wife, Catherine Holdsworth, was clearly a lady of considerable intellect who carried on her own correspondence with Newman. This led to her conversion to the Roman Catholic Church, and later to the conversion of most of their children, to William's distress. It would seem that, other than for religious matters, their marriage was a very happy one.

It is of interest that in describing the character of a man who achieved so much, his friends so often use the word 'idle'. It will be suggested that this is incorrect and that his problem was lack of concentration due to his exceptional range of interests. In his last years his friends saw him as over-working. It is almost inevitable for ease of understanding that this account of his life and work is divided by topic but this division does make it unclear how much was being carried on simultaneously; for one example, in 1857 he was upset by his wife's conversion but made the initial break in understanding rolling, whilst later work on rolling overlapped his early model tests, the building of his house and the problems of his children's conversion to Catholicism.

However, this great breadth of understanding was his greatest strength in technical advance, many individual aspects of his work were already known; his genius lay in putting everything together and applying the result to the solution of important problems.

A Note on Sources

When writing William Froude's obituary, (1) Henry Brunel took great pains to write to all those who remembered William but, even so, still complained of the lack of information on his early life, including his work for Isambard Kingdom Brunel on railways, which may excuse the present author's gaps in these areas. This obituary, which Henry said took half his working time over some months, is one of the most valuable sources.

There are three main sources for the early life of William Froude. Harper (2) draws mainly on conversations with his daughter, Eliza, known to the family as 'Izy', later the Baroness von Hugel, and with Mary Froude his grand daughter who also made available some letters by his brother, Hurrell Froude. The impression is that Harper was a careful researcher and that his account is accurate. However, it was written in 1933, Eliza died in 1931 and her reminiscences of William go back to his death in 1879, describing events of his and her early life. Mary was born in 1875 and can have had very little direct memory of her grand father.

L. E. Guiney's biography of Hurrell Froude (3) draws on '*The Remains of the Reverend Richard Hurrell Froude, MA, Fellow of Oriel*', published by Rivington in 1838 and on '*John Henry Newman, Letters and Correspondence to 1848*', published Longman in 1890. She also acknowledges the help of William's family and of the Reverend G. Kenworthy, Vicar of Bassenthwaite, on the Spedding family. The book is rather confused in presentation but seems accurate.

H. Paul (4) also acknowledges Eliza Froude who says that a large number of Anthony Froude's (brother) letters were destroyed on his death. Paul was writing in 1905 when Eliza's memories of William were more recent but there is no obvious discrepancy with Harper's much later account.

Abell (5), in his paper to the Devonshire Association, draws on much the same material but also acknowledges help from William Froude, the grandson, especially for help on Arnulph Mallock. There are a number of other references to the Froude family in papers to this Association, separately referenced, and thanks are due to the Secretary, Mr J. Bosanko, for making them available.

The main source for Froude's railway work and for a few other tasks up to 1859 is the '*Brunel Collection*' held in the Special Collection room of Bristol University Library. Few letters to or from Froude have survived. My thanks are due to Mr N. Lee and Mr M. Richardson for assistance in locating such letters.

The Brunel Collection also contains Henry Brunel's correspondence and from about 1859 this forms the principal source on the human background to his work. William Froude became an informal professional tutor to Henry on the death of his father, Isambard, and this relationship deepened both professionally and on a personal level. Henry acted as William's assistant in a number of projects and, as Henry matured, each found the advice of the other to be of great value. Henry used the phrase, 'almost a father to me' on more than one occasion, showing the warmth of their relationship. Henry's style of writing is colourful and reflects his mood. His descriptions of people, particularly those he did not like, are also highly coloured and should be treated with caution, while the author has had to struggle with the temptation to use too much.

Unfortunately, few letters earlier than 1872 from William survive. Mr G. E. Maby, former Archivist, has summarised the Henry Brunel letters and typed copies are held in Special Collections which are particularly valuable as Henry's letters are very fragile and many are almost illegible. Mr Maby and his son Mr R. G. Maby of Mabionics Ltd have computerised these summaries and have developed a search facility whose use I acknowledge with deep gratitude. Henry's letters, together with a few papers, notably to the Bath and West of England Agricultural Journal (6), are almost the only sources for Froude's activities during his long break from professional work, 1846–1857.

Froude's tutors at Oxford were his older brother, Hurrell, and John Henry Newman, later Cardinal, and William's long correspondence with Newman is invaluable in understanding how Froude's mind worked. There is much on the nature of proof and the differences between proof in religion and in science. These letters and the parallel correspondence between Newman and Mrs William Froude also deal with the emotional storms following the conversion of the latter and most of their children to the Catholic faith. I am most grateful to Mr G. Tracey, Archivist of the Oratory, Edgbaston, for making available these and some other relevant material.

There is a common problem with all these correspondence sources; William Froude, Isambard and Henry Brunel and John Henry Newman all had handwriting which is very difficult to read and there is an element of guesswork in some interpretations, particularly when read off faded 'press' copies. Froude and Newman sometimes seem to have written in haste and one particular letter of Froude's, dated 25 July 1864, which is more than usually difficult to read and is somewhat incoherent ends with a revealing postscript

– 'I have not had time to read and correct this – if it is obscure, do not waste time in trying to make out its meaning.'!

More usually, both took great care over their letters, marginal notes on the received letter showing plans for response. This correspondence seems to have been filed by Edmund Froude.

A Note on Sources

The files of the Public Record Office [ADM 136/117] are generally complete on the complex negotiations which led to Admiralty approval of funding for the new ship tank. As always, Henry Brunel adds background colour.

The building of the tank at Torquay is well covered in a large number of documents held in the Froudes' Museum in the Admiralty Research Establishment, Haslar, the direct descendant of the tank established in Torquay by William Froude and moved to its present site in 1887 by his son, Edmund. Until quite recent years, the establishment never threw away any document and there are 'In' and 'Out' correspondence files, invoices, diaries, as well as detailed technical records, drawings and reports.

When the new tank was set up at Haslar, Edmund formed a small collection of his father's apparatus which remained on display for many years. During World War II, the Director of Naval Construction, Sir Stanley Goodall, himself an ex Haslar man, arranged to transfer many of these items to the Science Museum. To mark the Centenary of the establishment in 1972 a large exhibition was held, mainly devoted to the then current activities. There was a small historical exhibition, organised by the present author, which included many of the bigger items held by the Science Museum and this proved so popular that it was developed into a permanent display as the Froudes' Museum. It covers not only the work of the two Froudes but also more recent items of historical importance. Access to the Museum may be obtained on application to the Director, ARE Haslar. The Museum is maintained as a spare time activity by enthusiasts on the staff and my thanks are due to Mr E. Maddick, Mr J. F. Anslow and Dr A. R. J. M. Lloyd in particular and to all who have helped in this work.

Both William and Edmund Froude believed in open publication and much of their technical work is very fully described in papers to the Institution of Naval Architects (it received the title 'Royal' to mark its centenary in 1960) and other engineering Institutions. While these papers have been quoted where relevant, there is no intention in this book to reproduce all his technical work which is readily available elsewhwere.

At the Sixth International Conference of Ship Tank Superintendents in 1951 it was decided to commemorate the pioneer work of William Froude in two ways; the erection of a plaque on the site of his original tank at Torquay and by the publication of a volume of his collected papers – proposed by Henry Brunel in 1880! This latter was published by the INA in 1955 and contains all but one or two of his papers (7). It does not contain the discussion on these papers of his nor, with two exceptions, does it contain the contributions which Froude made to the discussion of other papers. Both are important.

There are other contributions by Froude to Government committees on warship design and on scientific education, individually referenced.

Edmund Froude's successor, M. P. Payne read a paper to the INA in 1936 containing much previously unpublished work on the derivation of the Froudes' frictional data. His successor, R. W. L. Gawn, wrote a paper for the INA in 1941 describing the work of both Froudes at Torquay. Based on their reports, it is a valuable compilation whose value is increased by author's own experience in the same field of work. When he wrote this paper there were still men alive who had worked under William Froude and he acknowledges their help and contribution to the discussion of the paper.

Haslar was a very closed society and many men spent most of their working lives there. Edmund ruled for 40 years after which the next two Superintendents, Payne and Gawn, were in charge for 19 years each. There were men (females were not admitted until the late 1930s) at all levels in whom the sense of tradition was strong – in the 1960s Christmas presents were handed out at the party by 'Father Froude'. This led to a strong oral tradition which, where checks have been possible, has usually proved reliable. In a few cases where checks have not been possible, and the story seems likely to be true, I have used it but described it as legendary or 'it is said'. Many of these legends have been recorded in a paper by Gawn for his staff ca. 1954 '*Minore ad Majus.*' (Copy in Froudes' Museum). One persistent legend for which there is no direct evidence, is that the spectacular flying staircase in Froude's house was designed by Henry Brunel (Chapter 7).

The succesive Superintendents for the first 100 years have all shown interest, not only in advancing the technology which Froude created but also in recording the history of the establishment they directed. These are:

W. Froude	1872 – 1879
R. E. Froude	1879 – 1919
M. P. Payne, RCNC	1919 – 1938
R. W. L. Gawn, RCNC	1938 – 1957
R. N. Newton, RCNC	1957 – 1962
A. J. Vosper, RCNC	1962 – 1975
E. P. Lover, RCNC	1975 – 1981
R Burcher, RCNC	1981 – 1987
R. Spencer, RCNC	1988 – 1990
C.N. Stonehouse, RCNC	1990 – 1991

The title of Superintendant was not used again.

A Note on Sources

Notes and References

1. Obituary. William Froude. Proceedings Institution of civil Engineers, Vol LX, p 395, London, 1880. Written by Henry Brunel.
2. Gordon Harrington Harper. 'Cardinal Newman and William Froude – a Correspondence'. John Hopkins University, Baltimore 1933.
3. Louise Imogen Guiney. 'Hurrell Froude, Memoranda and Comments'. Methuen London 1904.
4. H. Paul. 'The Life of Froude.' (Anthony) Isac Pitman, London, 1905.
5. Professor Sir Westcott Abell. 'William Froude, His Life and Work.' Devonshire Association 1933 Vol LXV, p 43–76.
6. 'The Papers of William Froude, 1810–1879.' Institution of Naval Architects. London, 1955.

References to A Note on Sources

1. Professor Sir Westcott Abell. 'William Froude, His Life and Work'. Devonshire Association 1933 Vol LXV, p 43–76.
2. Louise Imogen Guiney. 'Hurrell Froude, Memoranda and Comments'. Methuen London 1904.
3. Herbert Paul. 'The Life of Froude'. (Anthony) Isaac Pitman. London 1905.
4. As 1 & 3.
5. As 1, quoting Keble.
6. As 3.
7. There is a book of water colours by Archdeacon Froude in the Exeter Cathedral Library (DC 35494) from ca 1816 and another owned by the Spedding family.
8. As 2.
9. As 2.
10. As 3.
11. As 1 & 2.
12. Letter by Margaret Froude held by the Oratory, Edgbaston.
13. As 3.
14. Piers Brendon. 'Hurrell Froude and the Oxford Movement', Paul Elek, London 1974. Quoting letter from Hurrell to William, 12.8.79.
15. Gordon Harrington Harper. 'Cardinal Newman and William Froude – a Correspondence'. John Hopkins University, Baltimore 1933.
16. Brian Martin. 'John Henry Newman – His Life and Work'. Chatto & Windus, London. 1982.
17. Rev T. Mozley. 'Reminiscences of the Oxford Movement'.
18. Oratory.
19. As 17.
20. As 15.
21. Mozley (17) says that William moved into Hurrell's vacant rooms in 1835. (Guiney says these were directly over Newman's in the chapel angle of the great quadrangle of Oriel College). This seems unlikely as the College at that time would usually only permit under graduates to live in. The dates given for William's dinners were taken from the records by the Archivist, Mrs E. Boardman, and passed to the author on 10.12.90.
21. Rev W. Harpley. Obituary Notice – William Froude. The Devonshire Association. 1879.
22. As 17.

Acknowledgements

Interest in William Froude is surprisingly wide and hence there are a considerable number of individuals who have helped and encouraged me in my work. Many archivists and librarians have been mentioned above including:

Messrs Maby, Lee & Richardson, Brunel Collection, Bristol; Mr G. Tracey, The Oratory, Edgbaston; E. Maddick, J. F. Anslow, Dr A. R. J. M. Lloyd, Froudes' Museum, ARE Haslar. To these one must add John Rosewarn, Secretary of the Royal Institution of Naval Architects; Mrs Boardman, Oriel College; Mr A. Holbrook, Bath University Library; C. Watson, Ministry of Defence, Bath; the librarians of Lambeth Palace and the Institution of Civil Engineers; the archivist of Westminster School; and the staff of Bath Public Library and the inter library loan service who obtained various books for me.

The work was made possible by the grant of a Rolt Fellowship by Bath University and this was due to Professor R. A. Buchanan, whose help and encouragement is most gratefully acknowledged. Professor D. R. Pattison of University College, London, has also been very helpful.

Mr L. Taylor and his late wife, Merlys, together with Mr G. D. C. Tudor have helped me from their own studies of Froude and Henry Brunel. Mr J. Spedding, descended from the family of Froude's mother, is thanked for his help.

Dr T. Wright of the Science Museum is thanked for permission to use his doctrinal thesis and for clearing up various points on apparatus held at the Science Museum.

Mr J. Lacy and Ms J. Cusack are thanked for advice and information on the yachting activities of both Froudes. The late Profesor Yoshioka, the only previous biographer of William Froude helped to inspire this work.

Kevin Gilligan, Anne Cowlin, Innes McCartney for help in the preparation of the book.

Finally, permission is gratefully acknowledged to publish extracts from books, papers, letters etc and to reproduce illustrations from: The Royal Institution of Naval Architects, The Institution of Civil Engineers, Bristol University, The Oratory, Edgbaston, The Froudes' Museum (Defence Research Agency), The Science Museum.

Chapter 1
William Froude – Family and Early Life

1.1   Family Background

The Froudes had lived in South Devon for several generations and were quite well off following the marriage of William's grandfather, Robert, to Phillis Hurrell in 1741. From the family tree (fig 1.1) it may be seen that the family name Hurrell is commemorated by the use of Hurrell as a first name in several generations. William's father Robert Hurrell Froude, was born at Wakeham House, Aveton Giffard, in 1771, a few months after the death of his own father (ref 1.1). Robert matriculated at Oriel College Oxford, in January 1788 at the age of 17. He obtained a pass degree in February 1792 and his MA in 1795. He took Holy Orders, became Rector of Denbury in 1798 and moved to Dartington as rector of both parishes in 1799. In 1820 he became, additionally, Archdeacon of Totnes, a post which he held until his death in 1859.

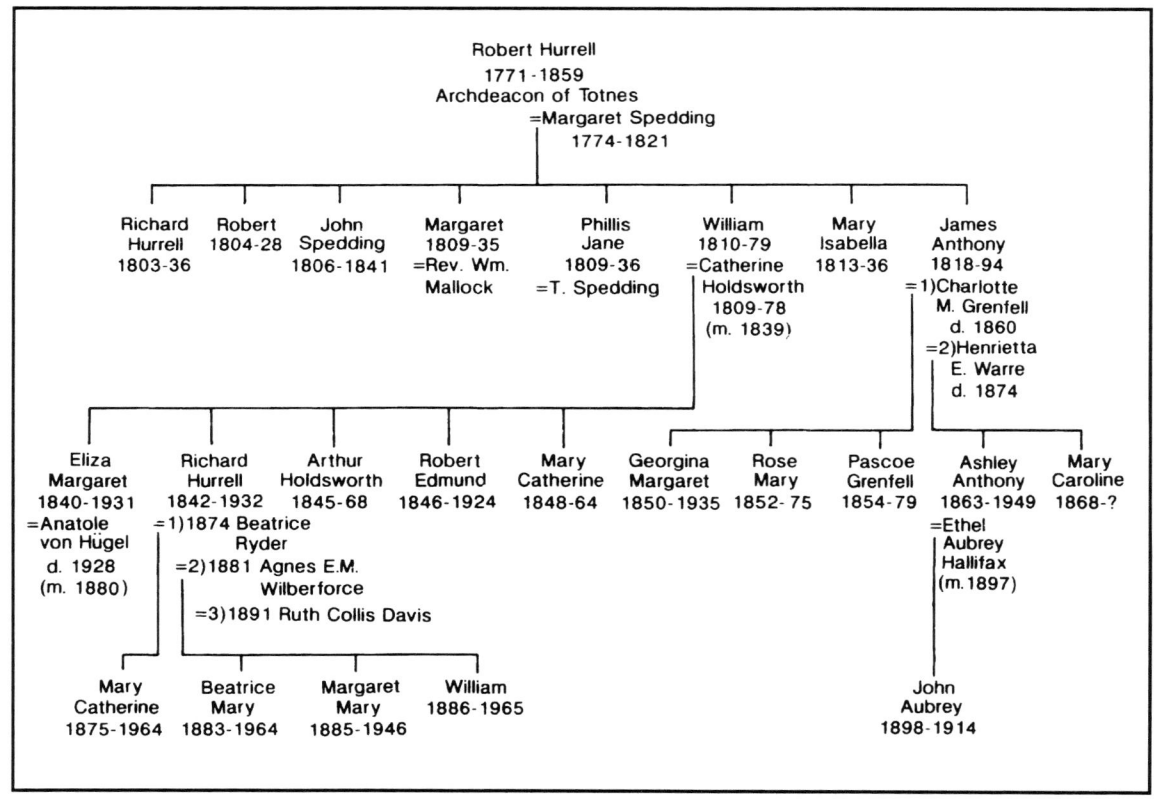

Figure 1.1 – Froude Family Tree

1.2 Archdeacon Froude

Both of William's parents were highly gifted and intelligent. Keble, the theologian, wrote of William's elder brother Hurrell, that he had inherited much of his ability from his father and his gifted mother.

> 'Like the one (father) he was clever, knowing, quick and handy; like the other, he was sensitive, intellectual and imaginative' (ref 1.2).

Keble continued of the Archdeacon,

> '...very amiable but provokingly intelligent, one quite uncomfortable to think of, making one ashamed of going gawking as one is wont to do about the world, without understanding anything one sees.' (ref 1.3)

At least on the surface, Archdeacon Froude was typical of the wealthy, land owning parson of the period with a private income of some £2–3000 per annum. He was passionately fond of hunting, a skilled and hard rider, a good judge of a horse and could afford to be the best mounted man in the field (ref 1.4). Robert was conscientious in all he did, farming fields with skill and knowledge and he was said to be the best magistrate in the South Hams. His advice was much sought, even to the extent of laying a troublesome ghost (ref 1.5).

His religion was deeply held and, though very conventional, he joined the Tractarians seeking reform of the Anglican Church. His youngest son, James Anthony, known later as an historian, wrote,

> 'My father was a High Churchman of the old school. The Church itself he regarded as part of the Constitution and the Prayer Book as an Act of Parliament, which only folly or disloyalty could quarrel with. ...the way to heaven was to turn to the right and go straight on.'

But it is worth noting that when his daughter-in-law, Catherine, became a Catholic, though deeply distressed, he neither disinherited William nor asked them to leave the Parsonage. He had a hatred of nonconformists, refusing to have a copy of 'Pilgrim's Progress' in the house. These views matched well with his strong Tory politics.

Much of what is known of the Archdeacon comes from his youngest son, Anthony, born five years after the last of his siblings, and who clearly had an unhappy childhood. Comparison of Anthony's view of his father with that of others who knew him suggests that, though not necessarily wrong, Anthony's description is certainly one sided.

Anthony wrote of his father:

> 'while in the diocese he gave the Bishop (the formidable Henry Philpotts, Bishop of Exeter, 1831–1869) always his steady support, as the authorised administrator of ecclesiastical land. He was sensible, strong, practical, with a fixed contempt for

eccentricities. He was perpetually occupied with business, but always silently, never I believe in his whole life appearing to make a speech on a platform, and despising and distrusting the whole race of popular orators. He never spoke even in private, of feeling or sentiment, and never showed any in word or action.'

Robert was not quite typical unless, as is possible, real achievement in some field was also typical of the wealthy parson. He was an artist of considerable merit and his sketches were compared by critics with those of the early Turner; he even won praise from that severe critic, Ruskin (ref 1.6), and several of his sons inherited some of his artistic talent. He read widely on current affairs and followed the work of explorers such as Franklin.

As a person he was described by Paul (ref 1.7), following Anthony, as 'a cold, hard, stern man, despising sentiment, reticent and self restrained' while Coleridge saw him as lacking imagination. Like many 19th century fathers he seems to have believed in the saying "Spare the rod and spoil the child". Certainly Anthony was well whipped. On the other hand, it is said that Hurrell's occasional differences with his father were over trivial points 'on such things as the culture of trees and the make and management of boats' (ref 1.8). Both Hurrell and his brother, Robert, shared their father's love of riding and were his frequent companions.

When Robert Froude came to Dartington Parsonage (fig 1.2) in 1799, he at once rebuilt the whole west wing, planted shrubs and vines and drained away the ponds. The eldest son, Hurrell, says in a letter of 1823 that '..it was situated in a steep and narrowish glen, which intersects a long line of coppice that overhangs the Dart for the length of nearly a mile, and rises almost perpendicularly out of the river to the height of about 200 feet. The stream there is still, clear and very deep; on the

Figure 1.2 – Dartington Parsonage

opposite side is Dartington and a line of long flat meadows, interspersed with large oak and ash trees, which forms the bank of the river. The steep woods on the Little Hempston side are in the form of a concave crescent.'

1.3　Margaret Spedding

Robert was at college with John Spedding of Mirehouse on Lake Bassenthwaite. The Speddings had a family tradition of interest in science, including mechanical engineering, as well as a great love of letters (ref 1.9). In 1902, when Robert was 31, he married John's sister, Margaret Spedding, then aged 27. She 'was very beautiful in person and delicate in constitution with a highly expressive countenance and gifted in intellect with the genius and imagination that the father failed in.' (ref 1.10) That such an attractive and intelligent lady as Margaret chose to marry Robert suggests that he was a more interesting person than that portrayed by Anthony. It seems to have been a happy marriage and fruitful one too, with five sons and three daughters.

Margaret had a great influence on her children, her eldest son Hurrell, attributed his religiousness, poetry and fire as well as his penetrating thought to his mother. Some idea of Margaret's philosophy and of the way in which she brought up her children may be gained from the following extracts from a letter to her second son, Robert, written about 1815, when he was aged eleven (ref 1.11).

> 'Thus God has decreed that the virtuous man only is to be the happy man.'

She goes on to explain that virtue consists of stretching both mind and body to the utmost saying.

> 'The acquisition of language in early life has been proved by experience to be the best means yet of strengthening the forces of the mind. This does not merely mean learning different names for the same thing but "voluntary self exertion", a real desire to improve.'

She continues that he didn't learn to swim by being driven into the water and whipped if he didn't float, but "You know by experience that you can only acquire strength and ease in the art by exerting your limbs and muscles to the utmost and by constant practise." ... "Ambition, riches, finery, vices may give a promise of happiness but such pleasures are a vanity and vexatious of spirit'.

She was not strong and died "of a decline", tuberculosis, on 16th February 1821, Margaret was sadly missed by her husband and children but Anthony's description written much later of a time of which he can have few memories since he was only three is almost certainly much exaggerated. He says that the Archdeacon retired into himself and nursed his grief in silence, melancholy, isolated, austere and never mentioned his wife's name again. The house was run by his daughter, Margaret, then not quite 13 year old, initially with some help from her aunt, a severe disciplinarian, according to Anthony (ref 1.12). Margaret was housekeeper for 23 years before she left in 1844 for a happy marriage with William Mallock, an important person in this account.

On the other hand, Hurrell (ref 1.13) describes a summer holiday in 1829 when the Archdeacon rented a large house at the mouth of the Dart and hired a 30 ton yacht. There was a party of 23, including the Champerdownes of Dartington Hall, and there was a lot of fun and games for all. They were all turned out of bed at half past five in the morning to bathe, great regularity was enforced at meal times and Mrs Champerdowne made them all read the psalms and lessons for the day before enjoying themselves with boating, fishing and more bathing. In the same letter, Hurrell wrote 'and Will, I suspect, will give up everything for the boats.' Two years later, Newman wrote

'When I was down at Dartington for the first time, in July 1831, I saw a number of young girls collected together, blooming and in high spirits and all went merry as a marriage bell...'

It does not sound as though the Archdeacon was a killjoy.

1.4 William's Schooldays

William, born 28th November 1810, was ten when his mother died and began his formal education at Buckfastleigh in a school kept by the Rector, Rev Lowndes, who took boarders. Anthony, writing of the school as he knew it, some seven year later, said that it was not a bad school for the period. There was plenty of caning but no bullying and Latin was well taught.

The older brothers, Hurrell and Robert had been sent to Eton but William joined Westminster School on 18th May 1823, as did his younger brother, Anthony, some years later. Again relying on Anthony's recollections of a slightly later era, he says that Westminster had a high reputation, the boarding houses were well managed, fagging in them was light and the moral tone was good.

In 1824 William was awarded a King's Scholarship which was very valuable as it not only provided for free education at Westminster but also gave privileged access to a closed scholarship to either Trinity (Cambridge) or Christchurch (Oxford). The scholarship was awarded on the basis of an extended series of oral examinations, known as the 'Challenge', which took place at the end of each day's school from January to May (ref 1.14).

William seems to have had difficulty in adjusting to life at Westminster as seen from Hurrell's letter of 28th October 1827 (ref 1.15).

> 'I am sorry to find that you are not on good terms with the other fellows in College, and though it is very possible, and indeed likely that the majority may in a case of that sort be in the wrong, still if it is only about amusements it is not worth while to set up for oneself on such trifles ..... If I was you I would try as much as I could to fall in with the general games, and even if it is a bore to be shinn'd at football, it is better than being disliked for shirking it. Besides it is no bad thing to be forced to bear pain now and then.'

Hurrell continues:

> 'I think I shall make rather a good teacher of Mathematics, and am very glad you like them well enough to read them by yourself. If I was you I would skip the 4th and 5th books, but read the 11th, also pay as much attention as ever you can to Greek plays, both to the scholarship and the construing, and then you will be able to start fair when you come here.'

> 'You ask me I see about the mathematical way of finding the way for a boat bow, now I fairly confess that I cannot tell myself, but I know this much, that till you are very familiar with algebra and fluxions you have not the remotest chance of following the process.

William is also accused of being careless with money, an allegation also made by Hurrell against Anthony later. There are hints that Hurrell saw signs of laziness in William but his views on his brothers must be treated with considerable caution. There were also well justified complaints concerning William's handwriting and demands for improvement which did not materialise.

Later, 10th February 1828, Hurrell wrote:

> 'I am very glad that Westminster air is beginning to set going again and to help up the leeway you had made at home.'

By this time, William's academic performance was already outstanding. In that year there was a process similar to the earlier Challenge in which the Dean of Christchurch and the Master of Trinity came down in state to the school to examine 'Major Candidates' orally in the 'Election'. Hurrell wrote on 10th February 1828

> 'I was surprised to hear that you ever had an chance of Christchurch and give you great credit for saying so little of it.'

Robert Froude, the second son, who had already gone up to Oriel and was known as a practical joker, died of tuberculosis on 28th June 1828. This further tragedy greatly upset William who appears to have decided that he had wasted his own life and must reform. This can only be deduced from Hurrell's replies to letters from William. (27th April 1828)

> 'From the tone of your letter you evidently seem in great distress at the recollection of your past life. ... For you know that one of the things with which we have always found fault has been your great closeness about your pursuits, so that we are quite in the

dark about you and hardly know you at all.'

By early May, William was more stable. Hurrell's letters make it clear that there was no dark secret for which William had to repent but the theme of idleness come up again on 5th February.

'As it is, I hope that the habits of idleness, though they will long be a great annoyance to you, may be opposed by resolution, and the trouble and self denial which this may take will strike at the root of selfish, dumpy ways, more than any agitation of feeling which you may have experienced.'

1.5   Oriel, Hurrell, Newman and Religion

William entered Oriel on 23rd October 1828. Hurrell writing to Newman on 12th August 1828. 'I have a brother who is coming to Oriel next term and will make a very good hand at Mathematics unless he is very idle.' He arrived at an exciting time. His brother, Hurrell, had become a Fellow in 1826 and, with Newman, had organised the election of Hawkins as Provost, following Copplestone's departure to become Bishop of Llandaff. In the re-organisation which followed, Newman and Durnford became senior tutors while Hurrell and Robert Wilberforce were junior tutors.

Hurrell and Newman had a new outlook on the role of tutors and were not content merely to lecture in class as had been the custom. They envisaged individual and group teaching which previously had only been available for payment. But in addition, as Newman wrote (ref 1.16): 'Secular education could be so conducted as to become a pastoral cure.' They realised that this approach would not be popular and kept Hawkins in ignorance of what they were doing. Inevitably he found out and from 1830 onwards acted to stop them taking on more pupils. It would seem that William learnt much of his mathematics from Hurrell.

Hurrell was now at the peak of his powers and was described by Mozley (ref 1.17), closely involved but very unreliable:

'As an undergraduate he waged a ruthless war against sophistry and loud talk, and he gibbeted one or two victims, labelling their sophisms with names... His figure and manner were such as to command the confidence and affection of those about him. Tall, erect, very thin, never resting or sparing himself, investigating and explaining with unwearied energy, incisive in his language and with a certain fiery force of look and tone, he seemed a sort of angelic presence to weaker natures. He slashed at the shams, phrases and disguises in which the lazy or pretentious veil their real ignorance or folly. His features readily expressed every varying mood of playfulness, sadness and awe. There were those about him who would rather writhe under his most cutting sarcasms than miss their part in the workings of his sympathy and genius.'

Hurrell was a born leader and admired by all who met him. Even his opponents praised his clarity of thought and his intellect. However, as Mozley hints above, there was also an unpleasant side to his character and even his mother saw him sometimes as a sadistic bully. William many years later, in early 1879, wrote to Newman suggesting that the latter had incorrectly described Hurrell in his writings. William wrote:

> 'Let me put before you leading points in which Hurrell differed from you. He had a masterly mechanical instinct and of mechanical philosophy, an impression with the dependence of human knowledge on human experience. He would have differed as to the practical logic of mechanics from the Grammar of Assent (ref 1.18).'

Williams's views on his brother's engineering skill must be given respect but it remains uncertain whether Hurrell's frequently declared belief in 'Probability as the Guide of Life' would have prevented him from following Newman to the Roman Doctrine.

Hurrell and Newman were now becoming close friends and allies in seeking to reform the Anglican Church. Newman paid his first visit to Dartington in July 1831 and seemed to enjoy the happy atmosphere and made a favourable impression on Archdeacon Froude, who was encouraged to support Newman and Hurrell in their reforming drive. He also met Catherine Holdsworth, later Mrs William Froude, and a regular correspondent of Newman.

During 1832 Hurrell developed tuberculosis and his father planned to take him on a Mediterranean cruise during the winter, inviting Newman to accompany them. It seems to have been a very enjoyable holiday, once early sea sickness had been overcome. Hurrell's health continued to deteriorate despite a holiday in the Barbados and when he returned in the autumn of 1835 Newman spent a month at Dartington saying farewell to his great friend. Hurrell died on 28th February 1836 and his father writing to Keble with the news said

> 'To me he was not only a most affectionate son but my companion and familiar friend, entering into my pursuits and amusements and guiding me by advice when I needed it.'

It was a sad time as Phillis died in 1835, shortly after her marriage, and Mary soon after Hurrell, all of tuberculosis.

All these events had a great influence on William, leaving him with a life long fascination with religious problems and an equally enduring friendship with Newman. William had apparently joined in the feuds at Oriel with an enthusiasm which he later came to regret. In August 1855 he wrote to the Provost, asking for his name to be removed from the Oriel Book and apologising to Hawkins. He continued

> 'I have often found myself vehemently and hotly advocating views opposed to those to which you were giving your support,

and perhaps joining in hotly in all the party denunciations to which such conflict of opinion is too apt to give birth. I now look back with wonder, and not without humiliation, at the heat and vehemence with which I joined in the fray, conscious as I ought to have been that I was taking part in a movement, not which my individual judgement imperiously drove me into, but which the exigencies of partisan warfare seemed to require.'

'I do not reproach myself for simply taking a side and giving votes on questions of which I knew I did not see the bottom or in a true sense of the word exercised my own judgement, but for taking a side with so much vehemence and heat; when although my real reason for taking it (indeed my avowed reason) was only that I gave in my adhesion to those good men who were its leaders. I could not but admit that many whom I know to be also good men were leaders on the other.'

1.6 Graduation

Clearly William's way of life at Oriel was not 'idle' as Hurrell wrote in 1831:

'W (William) continues very steady, getting up at half past five and working without wasting time until two or three.'

But the following year he was less certain –

'I am afraid poor William will make no hand of his Second Class. He has no interest and all his interleaves and margins are scribbled over with lug sails.'

William gained attention at Oriel for the depth of his scientific knowledge and Hurrell wrote that he 'over-reached the abilities of the younger dons.' There was a lighter side though; he dabbled in chemistry and his rooms were conspicuous by the long stains of acid running from his window to the ground two floors below and he was known as a practical joker, often involving the effects of laughing gas (ref 1.19). He is said (ref 1.20) to have contemplated a medical career.

It may be from this period that his skill with tools developed. Harpley (ref 1.21) says that he showed his friends a small locomotive engine which he had made himself and ran it on rails laid round his room. He also planned after graduating to return to Devon in a small river boat which he had strengthened and decked and fitted with a pump of his own design and manufacture (ref 1.22). Mozley was invited to accompany him but Froude discovered the latter's lack of seamanship and the trip was postponed though Mozley claims that the boat did finally reach Devon.

However, William achieved a First Class in Mathematics with a Third Class in Classics at Michaelmas, 1832. He did not entirely sever his connections with Oriel dining occasionally until 1834 (ref 1.23) and receiving his MA in May 1837.

References

1.1 Louise Imogen Guiney. *Hurrell Froude, Memoranda and Comments.* Methuen, London, 1904.

1.2 Quoted in 1.1.

1.3 Quoted in: Professor Sir Westcott Abel. *William Froude, his Life and Work.* Devonshire Association, 1933, Vol. LXV, pp.43-76.

1.4 Herbert Paul. *The Life of Froude.* (Anthony) Isaac Pitman, London, 1905.

1.5 As 1.4.

1.6 There is a book of water colours by Archdeacon Froude in the Exeter Cathedral Library (Ref DC 35494) from ca 1816 and there are watercolours in the possession of the Spedding family.

1.7 As 1.4.

1.8 As 1.1.

1.9 As 1.4.

1.10 As 1.3.

1.11 Letter by Margaret Froude (Spedding) held by the Oratory, Edgbaston. Undated but written ca 1816.

1.12 As 1.4.

1.13 Piers Brendon. *Hurrell Froude and the Oxford Movement.* Paul Elek, London, 1974. Quoting letter held in the Oratory from Hurrell to William, 12th August 1879.

1.14 Letter to the author from the Archivist, Westminster School, 10th June 1992.

1.15 Gordon Harrington Harper. *Cardinal Newman and William Froude – a Correspondence.* John Hopkins University, Baltimore, 1933.

1.16 Brian Martin. *John Henry Newman – His Life and Work.* Chatto and Windus, London 1982.

1.17 Rev. T. Mozley. *Reminiscences chiefly of Oriel College and the Oxford Movement.* London, 1882.

1.18 Oratory.

1.19 As 1.17.

1.20 As 1.15.

1.21 Rev. W. Harpley. Obituary Notice – William Froude. Devonshire Association. 1879.

1.22 As 1.17.

1.23 Mozley (1.17) says that William moved into Hurrell's vacant rooms in 1835. (Guiney says these were directly over Newman's in the college angle of the great quadrangle of Oriel college.) This seems unlikely as the College at that time would usually only permit undergraduates to live in. The dates for William's occasional dinners were taken from the records by the Archivist, Mrs E. Boardman, and sent to the author on 10th December 1990.

Chapter 1: William Froude – Family and Early Life

Annex 1.1 – Family Connections

The Froude Family

This annex is in no way intended as a family history and is intended only to provide background material which is needed to follow the life of William Froude. The first clear records show that the family was established in the South Hams, initially at Kingston, by the early 16th century (1). There was a great variety in the spelling of the family name, even into the 18th century though the line of descent is clear. (See family tree, Chapter 1.)

In 1746 Richard Hurrell, gentleman, of Modbury married Phillis Collings. (Note that this spelling of Phillis persisted in the family). Their daughter Phillis Hurrell married Robert ffroud of Walkhampton, third son of John, to whom descended the Modbury manors of Edmerston and Gutsford. They lived at Aveton Giffard where they are buried. Robert died four years after his wedding, leaving his widow with three daughters, Mary, Margaret and Elizabeth; their only son Robert Hurrell being born after his father's death. Phillis, described as a strong character (2) lived on for 66 years seeing several of her grandchildren achieve fame and many die young.

As described in Chapter 1, Robert Hurrell Froude and Margaret, née Spedding, had eight children of whom several were outstanding. The eldest, born 25 March 1803, Richard Hurrell is mentioned frequently in the text, however, since he was so prominent is his own brief career and was described by William as one of the three great influences on his life, some additional material is presented here. (See the book by Piers Brenden (3) for a full treatment.)

Hurrell was an inspiring leader, an innovative thinker on religious matters and a good and caring teacher. He was a leader of the reform group in the Church of England known as the 'Tractarian Movement' whose start is usually dated to 11 July 1833, when Keble preached at Oxford on the theme of 'National Apostasy'. It has been said of this group that 'Keble had given the inspiration, Froude (Hurrell) the impetus; then Newman took up the work.' The first 'Tract for the Times' appeared on 9 September of the same year. These Tracts were Anglican in tone, supported the 39 Articles and were intended to appeal to the clergy – Archdeacon Froude was a subscriber and supported the group.

During 1832, Hurrell developed tuberculosis and his father planned to take him on a Mediterranean cruise during the winter, inviting Newman to accompany them. It seems to have been a very enjoyable holiday once early sea sickness had been overcome. It is interesting that the Archdeacon and Hurrell, both experienced yachtsmen, suffered more than Newman.

Hurrell's health continued to deteriorate despite another

holiday in the Barbados and when he returned in the autumn of 1835 Newman spent a month at Dartington saying farewell to his great friend. Hurrell died on 28 February 1836 and his father writing to Keble with the news said

> 'To me he was not only a most affectionate son but my companion and familiar friend, entering into my pursuits and amusements and guiding me by advice when I needed it.'

There was a cruel streak to Hurrell, shown first in a letter written by his mother when she was ill in 1820 to an imaginary correspondent but really intended for Hurrell (4). She said that her son was '... very much disposed to find his amusement in teasing and vexing others ...' and continued that he disturbed her rest

> 'for what he calls "funny tormenting", without the slightest feeling, twenty times a day. At one time he kept one of his brothers screaming, from a sort of teasing play, for near an hour under my window. At another he acted a wolf to his baby brother, (Anthony, aged two) whom he had promised never to frighten again.'

After his death a two volume book *'Remains of the Rev Richard Hurrell Froude, MA, Fellow of Oriel'* was published in 1838 which most of his friends saw as a disaster; reflecting little credit on him. Guiney (5) quotes some comments on Hurrell justifying his mother's view of this fascinating but strange character, eg Rt Hon Sir James Stephen –

> '... very much disposed to find his own amusement in teasing and vexing others.'

The second son, Robert, died 28 April 1828, aged 24, '... full of frolic and power,' (2) Paul says (3) that he was a splendid athlete, compared by Anthony with a Greek statue, he had sweetness as well as depth of nature. His drawings of horses were the delight of the family and he wrote a graceful elegy for a favourite horse. The influence of his genial kindness was never forgotten by Anthony. Mozley (6) saw him as a practical joker, imitating the Provost of Oriel. Paul says of Hurrell and Robert

> '... striking, brilliant lads, popular at Eton, their father's companion in the hunting field and on the moors.'

The third son, John Spedding was a lawyer, reading for the bar when he died in 1841, aged 35. Anthony says that this brother would have preferred to have been an artist.

Margaret, the eldest daughter, ran the Parsonage until her marriage to the Rev William Mallock, Rector of Cheriton Bishop, in 1844, where her children were born. (See later note on Mallocks.)

Phillis married Thomas Story Spedding LLB in 1834 and died soon after at Dartington, on 14 June 1835. Guiney writes,

> 'Poor Phillis, the tradition of whose pathetic beauty yet lingers about the Cumberland fells wither she came as a bride'.

William came next, born 28 November 1810.

Chapter 1: William Froude – Family and Early Life

Mary was described by Newman,

> 'Mary Froude was one of the sweetest girls I ever saw. She was at this time engaged to Mr Bogue.'

(Rev Richard, son of a General). Hurrell called her Poll and used her to run errands.

> 'Mary Froude all the while was the very picture of naturalness and simplicity, receiving with equal readiness and equability the homage of one and the playful rudeness of the other.'

Her strength had been impaired by nursing Hurrell and she died 7 August 1836, aged 22, soon after marriage to her Richard.

Anthony was the youngest of the brilliant but tragic family. He was born on 23 April 1818, eight years younger than William. He wrote that after his mother's death he was ill treated at home by his father, Hurrell and Mary Spedding and only William was kind. However it is unlikely that Anthony saw much of William as he was away at school or University. After an unhappy time at Westminster, Anthony had a spell at home and then had a useless year with a tutor, finally arriving at Oxford in October 1836, four months after Hurrell's death.

The extracts from a letter which follows (7) from William to his father in October 1839 relate to an incident when Anthony had committed some serious offence at Oxford and William was going to Jersey to remove Anthony from bad company and either take him to stay with him at Bristol or leave him with Keble.

> 'I spoke to Jacob Ley – he said they had no reason whatever to imagine that this sort of offence had ever gone on before – Anthony's manner both when he was found and when he was brought up before the Vice Chancellor, did not in the least convey such an impression – he seemed very much embarrassed and overcome.'

> 'However, the first thing will be to get him away from the present position and company, and from what Dicky C says, I am not afraid that he will refuse to return with me. One thing which seems to dwell most in Anthony's mind and which probably encourages him in his want of openness is the notion that he is treated by us all as a child: and this notion makes him stuffy at home and break out into wildness elsewhere – perhaps a younger son **is** a little liable to be thus treated, and especially one who like Anthony has never done much in proof of manliness – and indeed one hardly knows how to treat him as a man.'

The text of the letter does not give much factual information but, to my mind, is from a son to a respected father and William seems confident that the Archdeacon will treat Anthony with understanding and forgiveness. It may be that Anthony exaggerates his difficulties with his father. It is also interesting that Anthony had become engaged in the Long Vacation of 1839 and is supposed to have given up his 'bad habits'.

Initially under Newman's influence, he wrote an essay on St

Neot which seems to have aroused doubts rather than faith. He was ordained Deacon in 1845. In 1847 he published, anonymously, two stories *'Shadow of the Clouds'*, which was almost an autobiography, and *'The Lieutenant's Daughter'*. He then became notorious for *'The Nemesis of Faith'*, published under his own name in 1849. It was burnt in College Hall by the Senior Tutor of Exeter, Rev W. Sewell. In 1856, the first two volumes of his great history were published. The volume dealing with the Spanish Armada was published in 1870 and includes a comparison with the mobility of steamships in which Anthony was assisted by William.

In 1863, he bought a house at Salcombe and began to see William frequently. (See Chapter 6 for Catherine's views of this friendship). Like his brothers, Anthony was a keen yachtsman.

He died on 10 October, 1894, the last of this generation of Froudes.

The Mallock Family

Margaret's husband, William Mallock, was a younger son of Roger, Squire and Vicar of Cockington Court, Torquay. Roger's wife, Mary, was a daughter of John Mudge, related to Thomas Mudge, a famous clockmaker, and it may well be wondered if Arnulph's mechanical skill was hereditary. After the death of Margaret's aunt, Margaret Froude, the Mallocks moved to Denbury Manor where the boys Arnulph and William Hurrell, (a writer 1849–1923) were educated mainly at home. William won the Newdigate poetry prize at Oxford and became well known as a writer. William Froude left £50 in his will to each of the Mallock brothers.

Henry Reginald Arnulph Mallock

After graduating from St Edmund's Hall, Oxford, Arnulph worked briefly with William Froude on the design and construction of the new ship tank and its apparatus at Torquay, (sited on land leased from his father) benefitting greatly from the intellectual atmosphere with Froude and his son, Henry Brunel, Beauchamp Tower, Baker and Metford. In 1876 he moved on to work for Rayleigh as an instrument designer for which he was to become famous (8). He later worked for a very wide range of organisations including the Admiralty, War Office and, later, the Air Force, reading the James Forrest lecture on Aerial Flight in 1912. Altogether he wrote 48 papers for the Proceedings of the Royal Society, becoming a fellow in 1903, and 89 letters to *Nature*.

He designed a vibration recorder for Edmund Froude which was first used in 1899 for the trials of HMS MEDUSA, (Chapter 14). This was found in a disused store in 1971, used by birds as a nesting box but, after washing, it still worked. As Boys said 'Mallock's apparatus was always simple but mechanically perfect, and it had the supreme merit that at least in his own

Chapter 1: William Froude – Family and Early Life

hands it worked accurately.'

The Spedding Family

The Speddings were originally Anglo-Irish, moving during the 16th century to Scotland, then early in James II reign to Cumberland. John Spedding and his wife Margaret owned Armathwaite Hall, in Bassenthwaite parish, Keswick, where their children were born. In order they were: Mary, Margaret (Mrs Froude b.1774), John, James, Anthony and William. When her father died Margaret was only seven. Armathwaite Hall was put in the hands of trustees but when John (1770–1851) came of age his patrimony was gone. He intended to join the Army, then in the Peninsular War. However, Thomas Story, a bachelor living at the head of Lake Bassenthwaite, made John the heir to Mirehouse, where the family still live (1993). Story died in 1802 so that John did not have to join the army but went to Oxford where he was Robert Froude's classmate (2).

John Spedding married Sarah Gibson of Newcastle, living to old age with many children. Their third son was James, a distinguished scholar, friend of Tennyson – "JS" of poem (3) – a writer on Bacon ("Impossible hero" JAF) and leader of the Cambridge set, "The Apostles". He was a close friend of the Froudes and there were frequent visits both ways. He introduced Anthony to Carlyle. The oldest brother, Thomas Story Spedding, married his cousin Phillis Froude (d. 1835).

Guiney says that the Speddings 'had imagination and a love of letters, and were ironic and opionative after another fashion. They had also, for generation after generation, as an unexpected corollary, a strong turn for science and even for mechanical science, as the less bookish Froudes, to offset their hard common sense, were restless and romantic lovers of the open air and of the sea ... All Hurrell's religiousness, all his poetry and fire and penetrative thought, came straight from his beautiful and highly intelligent mother, whom he lost just as he really came to know her, and whom he worshipped during the rest of his life. His stature, colour, and expression, as also, his delicacy of constitution, he received though her."

The Holdsworth Family

In 1839 William married Catherine Henrietta Elizabeth Holdsworth. Catherine's father was Arthur Howe Holdsworth of a very distinguished and wealthy Dartmouth family and her mother was born Henrietta Eastabrook. The Froudes and Holdsworths were long standing friends and there is a drawing by Archdeacon Froude in Exeter Cathedral Library, dated August 1817 of Widdicombe House. The wedding took place in Brixham Parish Church (Kingswear being in that Parish) since her father had quarrelled with J. H. Seale of Dartmouth where they might have been married.

Arthur Holdsworth had been Governor of Dartmouth Castle from 1809 and had been elected MP for Dartmouth in 1812, 1818, 1829 and 1831. Prior to the 1832 Reform Act Dartmouth had two MPs, Admiral Lord Howe being one from 1757 to 1782, and giving his name to Holdsworth. Catherine's father was interested in ships and took out a number of patents including:

1817	Gasometers
1831	Rudders
1840	Preserving wood.
1842	Constructing parts of ships to arrest fire and regulate temperature.
1846	Buoys and boat bouyancy
1849	Marine boilers and funnels of steam boats.

He was a shareholder in the Bristol and Exeter railway and may have met the Archdeacon and William as a result.

The Holdsworth family moved to Modbury from Astey in Yorkshire in 1620 when the Reverend Holdsworth became Vicar. The Lord of the Manor, Champerdowne, helped the Vicar's son, Arthur, in trade with Newfoundland which seems to have prospered. In 1650 Captain Arthur Holdsworth was directed by Cromwell to report on the state of the Newfoundland fisheries. By 1672 he was mayor of Dartmouth and had the title of 'Admiral of St Johns'.

The family continued to prosper, helped in 1725 by the award of "The Waters of Dart" from the Duchy of Cornwall in 1725 which entitled them to levy tolls on all goods landed between Salcombe and Torbay, a rich perquisite which lasted until 1860. The Holdsworths and their relations held most of the important posts in and around Dartmouth, Freemen, Mayors, Governor of the Castle from 1725, Rector of Stockenham and of Brixham, etc.

The family home was Widdicombe House, near Torcross, built in 1785 and enlarged in 1820. They also owned Brooke Hall, Dartmouth.

Notes on the Froude Family – References

1. Rev R. E. Hooppell, MA, LLD, DCL. *'The Froudes, or Frowdes, of Devon.'* Devonshire Association, 1892.
2. Louise Imogen Guiney. *'Hurrell Froude, Memoranda and Comments.'* Methuen London 1904.
3. Piers Brenden. *'Hurrell Froude and the Oxford Movement.'* Paul Elek, London, 1974.
4. As 2.
5. As 2.
6. Herbert Paul. *'The Life of Froude.'* (Anthony) Isaac Pitman. London 1905.
7. Rev. T. Mozley. *'Reminiscences chiefly of Oriel college and the Oxford Movement.'* Longman Green, London, 1882.
8. Letter in Lambeth Palace Library, Ref MS 168 ff 63–4.
9. C. V. Boys. *'Obituary – H R A Mallock'*. Proc. Royal Society, 1933.

Chapter 2
Railway Engineer

Froude joined H. R. Palmer as a pupil early in 1833, working on the Parliamentary survey for the South Eastern railway (ref 2.1). Palmer was a distinguished civil engineer of the day who had been Chief Engineer to the London Docks and was later to become President of the Institution of Civil Engineers. The choice of route for the South Eastern Railway ran into serious problems and Froude must have learnt of the need to soothe both Parliamentary and local opposition. The survey was completed at the end of 1835 and Parliamentary assent was given in 1836 with a cost estimated at £1.4 million.

Palmer had carried out experiments in 1824 using a spring balance to measure the resistance of barges on the Grand Junction Canal (ref 2.2) and, in the light of Froude's already keen interest in the performance of ships, it seems possible that he discussed this work with Palmer.

In 1837 Froude joined I. K. Brunel's engineering staff, working under William Gravatt, Chief Assistant for the Bristol and Exeter Railway, an appointment which may have been eased by the fact that the Archdeacon was a considerable shareholder in the line. A. H. Holdsworth, an old family friend, who became Froude's father in law when he married Catherine in 1839 was another big shareholder. Understandably, Froude's duties over the next four years remain obscure as might be expected for the work of a very junior engineer. Brunel seems to have worked his assistants hard but defended them from outside attack and treated honest claims for expenses generously (ref 2.3).

During these years Froude developed a new approach to the design of masonry bridges crossing on the skew which involved the mechanically correct principal of spiral tapering courses with the joints in every part of the arch being made at right angles to the lines of pressure. This makes a much stronger structure than the more normal design with spiral parallel courses in which much of the strength depends on friction between courses at the joints. Unfortunately, no details of this work of Froude's have survived but this early success, which was used in at least two skew bridges on the Exeter line by Froude must have won favourable notice from his employer. These bridges were close to Exeter at Cowley Bridge and Rewe (fig 2.1) (ref 2.4). It is believed that Brunel used the design on other lines, particularly the Gloucester line near Berkeley.

Round about 1839, Froude worked on the curvature of the track required to minimise the jolt due to sideways forces on trains entering a bend. Summarising his later paper very briefly, Froude began by saying that, in a curve the outer rail was elevated above the inner by an amount known as cant, which varied inversely as the radius.

Figure 2.1 – The railway bridge at Rewe showing the skewed brickwork pioneered by Froude

The problem came if the radius of the curve altered suddenly, implying a step in the cant, the usual solution being, as Froude put it, 'to humour it in', a problem familiar to operators of model railways with a limited range of curves. When the curvature reduced, a short straight was inserted 'to make things pleasant.' Froude said that curves of different radii should not meet but should be merged by a gradual change of curvature. In a mathematical annex, he showed that this curve should be close to the shape of a cubic parabola, the mathematics being similar to that governing the bending of a cantilever, loaded at the end. If the difference in cant over the length of a locomotive was large, the frame could twist if there was insufficient movement in the springs.

For reasons which are unclear this work, which he referred to as a "curve of adjustment" was only published in 1860, (ref 2.5) though the results were common knowledge long before. It is interesting that in the discussion, Gravatt, for whom Froude was working in 1839, claimed to have found an alternative solution which was more correct though less easy to use than Froude's method.

By 1841 Brunel was becoming increasingly dissatisfied with progress on the Bristol and Exeter and by April he was writing to Gravatt concerning changes in management. (6th April 1841) (ref 2.6) 'Froude is the only one (of the assistants) that I have made up my mind to keep on as yet....' Before long, on 6th June, Gravatt, himself was being criticised for failing to inform Brunel of what was going on. 'I have told Froude that I look to him individually for his report of certain observations of level which will help to get centring agreed. A few days later, on the 18th, Brunel told Gravatt that he had lost confidence in him and

Chapter 2: Railway Engineer

Gravatt resigned shortly afterwards, but would have a distinguished career elsewhere as an engineer. Froude was to stay as Brunel said in his letter to Gravatt. 'I told you at the same time that you might inform Froude that he could certainly remain.' and the assistants should all go 'with the single exception of Froude.'

In November Froude took over management of the line from White Ball tunnel and immediately showed his spirit writing to Brunel:

'Cullompton, Nov 26, 1841.

My Dear Sir,

I fear that it will seem officious in one that having been sent down yesterday with instructions to commence operations on this district, I should today set out with writing to you on a matter but little concerned with it, but when you have looked over what I send, I trust you will excuse it and at least give me credit for good intentions. ....'

He goes on to say that while looking for a missing cross section he came across some of the original trial borings which led him to propose a new summit route with a smaller cutting.

'I trust you will forgive me for having undertaken a task which might seem to be out of my province. I have mentioned the matter to you alone (except my Father, whose advice I have asked about it and I think I may fairly urge as an excuse for my anxiety to have the question examined, the fact that my Father holds shares to a considerable extent and is of course proportionately interested in the possibility of a diminished expenditure.) I expect you will be down this time tomorrow.'

At this time the newly married William Froudes were living in Cullompton where he seems to have had a very happy time and made many friends. In June 1844, he was presented with a silver salver 'in token of respect for his charity, benevolence and usefulness during his residence in the town of Cullompton.' (ref 2.7)

A paper on the Church of St Andrew, Cullompton, (ref 2.8) says that the chancel roof was restored and recoloured gilt in 1849, half the expense being borne by the Vicar and half "by a casual resident, Wm Froude, who inserted the iron stringers to prevent the clerestory roof from spreading." The writer says that Froude had an exaggerated notion of the vibration likely to be caused by the passage of trains, which seems unlikely since the church is about a quarter of a mile from the line. Since Froude was not living in Cullompton in 1849, this story may be wrongly dated.

In 1842 Froude seems to have worried about his next job and spoke on the subject to Brunel who replied on 20th November to the effect that the future was unclear.

'As regards the Plymouth line it is very difficult to say what my present ideas are as I really cannot see ahead. I never intended until later to take any steps myself but I think it is not impossible that steps may be taken by others which would necessarily lead to my being engaged on it. .... I think in all probability there will be a shuffling of cards and a struggle between engineers, solicitors and interested people in which I shall unavoidably be mixed up from my commitments with other companies.'

However, Froude stayed on the Bristol and Exeter, even taking full charge during Brunel's absence as the latter wrote to J. B. Badden on 21st April 1843. 'I shall send full directions to Froude to prevent the works being delayed during what I trust will be a temporary absence...'. Brunel's increasing trust in Froude is shown by a letter to W. Brockenden from much the same time, 'I have not the slightest objection to your son going to Froude's but on the contrary should be most happy to give him every encouragement and assistance'. This task came to an end when the line to Exeter completed on 1st May 1844, the Directors' Special leaving Paddington at 07:31 and reaching Exeter at 17:45. The latter stages of Froude's railway career are not easy to follow and key events, as far as they are known, are set out below.

1844 May	Bristol and Exeter line completes to Exeter.
1844 Autumn	Survey on West Somerset.
1844 Sept	Survey on South Devon (Atmospheric) Line.
1844 Late – mid 1845	Sick leave.
1845 Late	North Devon Survey.
1846	Cornwall line with Johnson.
1847 Jan	Pumping engines started on S. Devon.
1847 early	Correspondence with Brunel over paper on compressible fluids leading to tests in support of this paper using part of atmospheric line. These probably led to his patent on an improved seal for the atmospheric line.
1848 Jan	Patent taken out.

In the autumn of 1844, Froude was in charge of the Parliamentary survey of the West Somerset and Weymouth line. On 23rd December, Froude received a formal note enclosing a cheque for £612-7-8 as salary for his work on the survey and from 30th September on the South Devon railway. It was a very high salary by Brunel's standards and indicates that Froude was now a senior member of his staff.

It seems Froude was ill in the Autumn of 1844 as the following extracts from a letter from Brunel show. Brunel had not answered Froude's letter of 12th May and had

'... hoped for a better account of yourself. If, for six months, you really must take the care of yourself which you speak of, it would be wrong of you to attempt during that time to take the superintendence of works which will require a great deal of

Chapter 2: Railway Engineer

Posting...'

There is then a reference to the great inconvenience to the Company.

> 'I am doing you a great service by taking on my self the principal duty of deciding for you......There will be plenty doing when you come back. Make a resolution as you would a religious vow – if you ever make one, to have nothing to do with business except such as I may trouble you with.'

A very long, kind letter, which shows that Froude is very highly regarded. There is no clear evidence what this illness was but, in a letter to Catherine dated 28th May 1844, Newman asks 'How are William's eyes.' (ref 2.9) In later years his health was not good and there is at least one reference in Henry Brunel's letters to eye trouble.

On 22nd July, 1845 I. K. Brunel tempted him back to work:

> 'The N. Devon project is seriously going in the company formed etc – I suppose I cannot calculate on you to superintend in any way the survey – if I could I would place such a surveyor under you as would relieve you from any labour but a *gentlemanly that is idle sort of surveillance* on the supposition that this [would be impossible]. I offered it to England – he has declined on account of the uncertainty of present engagements.
>
> 'If I put it under the charge of Peniston, I wish to know from you the names of your assistants and their qualities. If by any change of plan you should be disposed to **accede** yourself with the **survey** during the autumn merely to direct if not to attend to plans or standing orders let me know as it would be a great relief to me.'

(Notes: Underlining is Brunel's. Italics the author's. The phrase 'gentlemanly that is idle' is of interest, suggesting that Brunel already saw Froude as a bit of a dilettante, a topic which will be further considered in the last chapter.)

Froude seems to have replied by return accepting the post as Brunel's next letter on 26th July shows.

> 'Your letter has been an immense relief to my mind and I feel under greater obligation than I should be willing to place myself under to anybody else but you. I have been compelled to the destruction of my comfort to undertake a great deal more than I can possibly attend to with credit or satisfaction to myself – and but for your help on the North Devon I should have dropped a huge stitch in my work.
>
> 'As regards the work that you will have – I will give you a first [class] Master Surveyor who will take the whole trouble and all the details, direct preparation of plans etc comfortably off your hands. I really believe that you will have little else to do than to ride about...'

The letter concludes with arrangements for a meeting. Froude was to be paid £5-0-0 per day plus expenses, a very high rate indeed for Brunel to offer.

This work soon ran into difficulty (ref 2.10). The unhappy story begins in 1838 when the Taw Vale Railway & Dock Co was incorporated to build docks at Fremington and a railway thence to Barnstaple. Nothing happened until in 1845 the Company revived its powers with a view to selling itself to the proposed North Devon Railway, which with the support of GWR, Bristol & Exeter and South Devon was to continue the proposed broad-gauge route of the Exeter and Crediton on to Barnstaple. The Exeter & Crediton obtained its Act in 1845. (An earlier Exeter & Crediton of 1832 came to nothing). The North Devon line was too late for Parliament that year, and having refused to buy the Taw Vale found itself opposed by a Taw Vale Extension standard gauge scheme (with LSWR backing) in the 1846 session.

The North Devon failed Standing Orders, but the extension had problems too as local people wanted to avoid a change of gauge at Exeter. They obtained an Act in August leaving open the gauge question. There were further problems and the broad gauge only reached Barnstaple in August 1854. Froude blamed himself for this failure, presumably for not recognising the political problems in getting Parliamentary approval. He wrote to Brunel offering to forego his salary, eventually receiving the following reply, dated 19th December 1845.

'My Dear Sir,

I have just recollected that I have not answered your two letters which reached me at Exeter. I did not mean to answer them immediately but did not intend to delay so long. Your idea about not receiving any salary – however excellent may have been the feelings which gave rise to it won't do at all in practice unless you and I are both prepared to avow publicly and to the Directors of the company that there has been gross negligence and you are the principal if not the only party to blame – now even if this were true it would hardly be to the interests of the Company to avow it and if not true it would certainly be very prejudicial to their interests to do anything which would give a good foundation for the report. No – everything must be done soberly and in the ordinary course for the sake of appearances even if for nothing else – but even as between ourselves it would be quite wrong and quite absurd for me to allow any such feeling as you have expressed to be acted upon – you did your best and the utmost I can say now that I am no longer afraid of annoying you is that you made a great mistake in not perceiving the danger sooner – quite a strange unaccountable mistake but from that very circumstance it is one of those which no one could impute to anything but a very singular accident – it's very unfortunate and that is all that one can say – you must have the goodness therefore to send me the statement for your own salary. I understand you are to be in Town after Christmas Day and I can then talk to you on other matters.

'My great expectation is that the Bristol & Exeter dissenting

Chapter 2: Railway Engineer

> proprietors (not including the Archdeacon) would impute to the failure to materialise on my part has been but very partially realised and I think the report which did just rise to the top has died a natural death.
> Yours very truly,'

On 3rd October, 1846, Brunel wrote again asking for help on the Cornwall line for which an elderly but very competent surveyor called Johnson would do the bulk of the work:

> 'but I want above that again a general who can judge of local interests etc. etc. .... It is I am sorry to say not by any means in conjunction with the B. & E. who have elected to be at war with the GWR have accepted my resignation as their Engineer.'

It is not certain if Froude accepted this task though he probably did since his long friendship with P. J. Margary, the Chief Engineer of the Cornwall Railway, seems to date from this time, but it is a nice tribute to his tact and diplomacy. By the end of the year Froude had retired to Dartington to help his father who suffered from a recurring illness until his death 13 years later.

Though this marked the end of Froude's direct work on railways, he was consulted quite frequently in later years by Henry Brunel, Isambard's son eg South Hams Railway (1865), Brighton (1866) and the Chelston line (1865).

Froude had one more contribution to make to railway engineering, in a paper read to the Institution of Civil Engineers in 1847 dealing with the flow of compressible fluids through orifices (ref 2.11). He tells us that his original intention was to consider the action of steam throughout the locomotive engine. However, he found that the discharge through the blast pipe alone was a sufficient problem. The first draft did not go very well as on 7th April 1847 Brunel wrote to him in more than usually illegible handwriting.

> 'I deny altogether the very foundation of your theory now that you lay it bare – the rate of expansion will not be infinite in the case you assume nor in any other besides which it seems to me that you are making a physical mistake or else a mathematical one in your statement that the velocity of the ——— particle is equal to the velocity of the unexpanded particle.'

(Some words are quite illegible and are replaced by —)

> 'By the rate of expansion by particle I suppose you mean ultimate particle which may be more or less closely in contact but themselves are supposed to remain unchanged – now the velocity of an ——— particle will be equal only to the sum of the velocity of the particle about to ——— and the velocity of expansion or separation of the particles not their product.'

> 'This seems to me all very evident and you are evidently very wrong but it is so rare a thing for me now to work a mathematical question that I feel I may be altogether writing nonsense – one sadly loses the habits of mathematical reasoning – the subject is

of great importance to me just at present and I should like you to pursue it.'

One could only write such a rude letter to a very close friend!

The final version of the paper opens with a robust declaration '...the law of discharge usually assumed is erroneous;..'. For the first but by no means the last time Froude is overturning established views. In this case he realised that previous work had assumed that laws which apply to incompressible liquids also apply to gases, something which is correct at low speed but ceases to be true as the speed of sound is approached. Having derived some fairly straightforward mathematical expressions, he then used some data supplied by Daniel Gooch on the engine IXION and showed that the steam port absorbed some 50 of the 320 horsepower available.

Mention has already been made of his brief survey work for the South Devon Atmospheric Railway in September 1844 and though this appears to have been his only direct involvement with that line as a member of Brunel's staff, he kept in touch with the progress of the line and was eventually to write that chapter in the Life of Isambard Brunel, compiled by his sons.

Pumping stations at the Exeter end of the South Devon railway began to work from January 1847 and by the autumn the line was working as far as Teignmouth. Later that year, Froude carried out tests using the atmospheric railway in support of his paper on flow of compressible gases. He acknowledges the help of all concerned but, in particular, that of G. D. Kittoe who prepared the apparatus and assisted in running the trial. Kittoe was to carry out similar work for Froude for the rest of his life. At this time Kittoe was working for Brunel but in 1857 he became dissatisfied with Kittoe's work which was strongly defended by Froude in several letters concluding in one dated 9th August 1857 that he was '...very glad to find that you have yourself seen reason to judge that Kittoe's shortcomings have been due to a greater pressure of circumstance than you had reason to think previously.'

Quite early in the operation of the atmospheric railway problems were experienced with the seal which held the vacuum in the tube of the railway. The seal was a leather one and the problem was not that the leather was eaten by cows as is often stated. Hadfield (ref 2.12) suggests that rust was the main cause, the pipes had been allowed to rust before installation and this was aggravated by the salt laden air near the sea. Other atmospheric railways used beeswax and tallow on the seal which may have preserved it whilst the South Devon used lime soap which may have accelerated rusting. Rust would attack the leather, causing it to tear.

Froude, perhaps as a result of the tests described above, designed an improved seal for which he took out a patent on 5th January 1848 (12014 of 1848). This used a pair of metal flaps in

Chapter 2: Railway Engineer

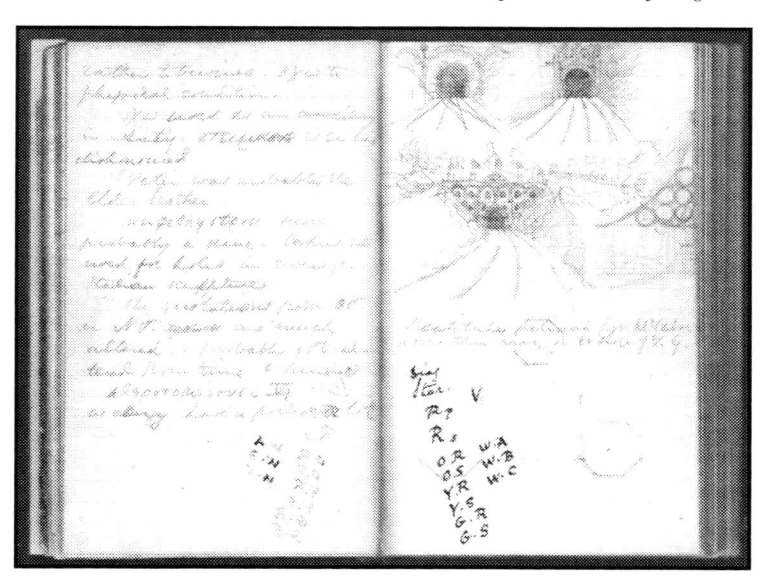

Figure 2.2 – Pages from Foude's notebook of his railway days

short lengths, giving strength, backed by a long strip of vulcanised canvas which formed both the hinge and the seal. After the driving piston passed, the sealing strip and the flap would be pressed down onto the seat compressing the rubber and forming a tight joint. There is mention of an improved rubber seal being tried near Dawlish just before the atmospheric system was abandoned in August 1848. It is tempting to think that this may have been Froude's design though there is no direct evidence.

While Froude's railway work is over-shadowed by that of his mentor, I. K. Brunel, his achievements by the age of 36 are ahead of most young engineers. He had supervised, successfully, a major project for Brunel, carried out several surveys, written or drafted papers on the curvature of track and the expansion of steam, designed a novel and effective brick bridge and patented a new seal for the atmospheric railway. The very wide range of his talent is already apparent.

References

2.1 Professor Sir Westcott Abell. William Froude, his Life and Work. Devonshire Association, 1933, Vol. 65 pp.43-76.

2.2 H. R. Palmer. Experiments on the Resistance of Barges on Canals. Trans. Inst. Civil Eng, Vol. I, pp.165–173. 1824.

2.3 Sir Alfred Pugsley. I. K. Brunel. Engineer. p.13. ICE, London, 1976.

2.4 In preparing Froude's obituary, Henry Brunel discusses possible sites for these skew bridges in a letter to Edmund Froude. (28232 100180). One possible site was on the London side of the Down distant signal between Stoke Cannon and Rewe. SX 945994. The other was Cowley Bridge on the A377 at SX 909953.

2.5 W. Froude. On the Junction of Railway Curves at Transitions of Curvature. Trans. Inst. Engineers and Shipbuilder in Scotland, Glasgow, 1860. Note that this paper is not in the collected papers published by the INA.

2.6 Brunel Collection, Bristol University Library. Dated letters are not separately referenced.

2.7 As 2.1.

2.8 Rev. Edwin S. Chalk. The Church of St. Andrew, Cullompton. Trans. Devonshire Ass. Vol. XLII, 1910, p.185. [Leslie Name of paper and author]

2.9 The letters exchanged between Froude and Newman usually conclude with a line or two concerning their health. Letters in this period are scarce but there is no mention of illness.

2.10 D. St. J. Thomas. West Country Railway History. David and Charles, Newton Abbott, 1960, p.92.

2.11 Froude. On the Law which governs the Discharge of Elastic Fluids under pressure, through short Tubes or Orifices. Trans. Inst. Civil Eng, 1847. Reprinted in collected papers, p.1.

2.12 Hadfield. Atmospheric Railways. Alan Sutton, Gloucester, 1985.

Chapter 3
Early Retirement

3.1　The Lotus Years 1846–1857

As mentioned earlier, William Froude gave up full time professional work in 1846, to help his father who was ill. Froude later wrote to Isambard Brunel on 8th May 1857

> '...until my Father's regular spring illness had settled in on him; and this concurring with general external circumstances of interruption, over which (as the phrase is) I had no control had so shortened and broken up my time...'

In a well off household, looking after Father can hardly have been a major preoccupation and it is probable that William's help was more in the nature of managing his father's estate. The other 'interruption' referred to is presumably his wife's conversion to the Catholic faith, referred to a little later.

3.2　Marriage

In 1839 William Froude married Catherine Holdsworth of Widdicombe, near Kingsbridge. Her father, Arthur Howe Holdsworth, had been Governor of Dartmouth Castle since 1809 and had also been MP for Dartmouth until the Reform Act of 1832. Her grandfather, too, had been Dartmouth's MP and her great grandfather had also been Governor of the Castle. Catherine's father, A. H. Holdsworth was an inventor and it is interesting to note that 5 of his 6 patents relate to ships. Catherine's mother was Henrietta R. Eastabrooke (ref 3.1). Catherine's background was clearly comfortable, secure, authoritative and inventive.

Though there were to be serious difficulties over religion, it is clear that the marriage was a happy one and mutual affection did not fade. (See Chapter 15) By the time William retired in 1846 there were four children; Eliza (b 1840), known to the family as 'Izy', (ref 3.2) Hurrell (b 1842), Arthur (b 1845) and Edmund (b 1846) and a fifth, Mary, came in 1848.

3.3　Gentlemanly Activities

When William retired in 1846 he lived the life of an active and responsible country gentleman. He was a Justice of the Peace; he concluded the letter to Isambard, quoted above

> 'Since I began this letter I have had a long interlude of arduous magistrates business – and a good deal of work for my father.'

He was a most conscientious trustee for several turnpikes;

there is a letter, a full two pages long, dated 22nd February 1878, to a defaulting ratepayer

> '...we desire to represent to you in a friendly manner that it is very much to be regretted that you have refused to pay in full the highway rate of four pence in the pound...'

Froude was a member of the Dartmouth Harbour Commissioners, a director of the Dartmouth Steam Packet company as well as the owner of the paddle steamer "Royal Dart", so named because Queen Victoria had once taken a short river trip in her. He was involved in the design of sea defences for exposed lengths of Torbay and the curved forms which he used seem to have stood the test of time.

3.4　　Farm machinery

William Froude acted as a judge of machinery at agricultural shows and wrote several papers which show his determination to measure and analyse rather than rely on subjective judgement, which was to become such a feature of his work. In the first of these papers, in 1857 (ref 3.3), he began with a lengthy explanation of the technical meaning of "power", the rate of doing work. Then, as now, the familiar unit was Watt's "Horsepower" of 33,000 ft. lbs. per minute (ie lifting 33,000 pounds through one foot in one minute). The horse which Watt chose for comparison was a puny one so that comparisons with the steam engine would be favourable.

Froude explained that a good, average West Country plough horse, pulling 180 pounds for an eight hour day at two and a half miles an hour would average 1.2 horsepower. He then showed that the horse's power would fall off if its speed exceeded two and a half miles per hour or fell below two miles per hour. In an aside, he showed that a stage coach would need a pull of about 1/30 of its weight on a good, level turnpike which meant 65 pounds from each horse or 80–90 pounds allowing for gradients. They would cover about 9 miles in an hour which would leave them exhausted for the day – it was said to be a "killing pace" – even though the work done was only one third of that of the plough horse. He also introduced the concept of the difference between the power supplied and that which performs useful work, so describing the efficiency of the machine.

That paper went on to discuss the theory and geometry of measuring power from the difference in the tension in the driving belt either side of the driving pulley. Another pulley would move against a spring, recording the force automatically on a rotating drum. An oil filled damper prevented undue oscillation in the records. All these features would re-appear in later Froude dynamometers.

Chapter 3: Early Retirement

In his 1858 paper (ref 3.4), he described the mechanical design of such a dynamometer and, in an example, showed that threshing (he spelt it 'thrashing') wheat needed much more power than barley. He went on to discuss the principles of a friction brake dynamometer invented by John Imray of Lambeth, giving the results of many tests on belt drive.

The next agricultural paper (ref 3.5) critically reviewed the farm machinery on show at Truro in 1862. It began with a careful analysis of the movements and actions of reaping and mowing machines comparing the performance of different designs and this is followed by a similar treatment of threshing machines. The design of wire fences is dealt with in detail; the size and shape of posts are considered in relation to the loading from the wires and the support from the soil. As a result, Froude notes that he used six inch square posts instead of the usual eight inch but with wedge shaped attachments to give the base a better grip in the soil. He took the load of a cow leaning on the fence as four hundredweight which, at normal post spacing, would deflect the wire sufficiently to break it rather than the post.

Froude then turned to the steam engines used to power the threshing machines. Usually, three horses were used to tow the engine and three more for the threshing machine and the cost of the horses, their drivers and that of the engine man's wasted time was considerable. He suggested that the engine should be used to drive wheels so that it could act as a tractor to tow the machine. A Mr. Taplin had a prototype along these lines but Froude thought it unsatisfactory; a horse was used to steer it and the engine weighed six tons. He was already in touch with Beauchamp Tower on light weight steam engines and, with Henry Brunel, was dreaming of their possible use in a flying machine.

Froude's last agricultural paper was not written until 1869 but is included here to complete the topic (ref 3.6). In this paper he reviewed the machinery on display at Falmouth, returning to the theme of lightweight steam engines. He complained that the 'portable' engine of eight horsepower weighed about 60 hundredweight (cwt) and this should be cut to about 10 if it was to be truly portable in Cornish lanes. He approved of an engine intended for steam launches made by Plenty of Newbury which weighed only 11 cwt for 8 HP (about 18 cwt in a wheeled carriage). Plenty had recognised the importance of providing a generous heating surface and of building in steel rather than in iron. He had experience with a Plenty engine in a launch lent to him by a friend (Bidder) and found it reliable (Chapter 9). Froude then showed that further reduction in weight could be achieved by increasing steam pressure and working at higher speed.

He also showed that fuel savings could be made by supplying

the boiler with hot feed water but pointed out that the feed pump should work on the cold water before it was heated, to ensure that the hot water did not boil under the reduced pressure in the pump. Howard's boiler met with his approval but the firm was castigated for failure to put test results in the brochure or to supply them to their representative at the show – still a common problem today. Figures which he obtained later showed that it could boil nine pound of water per pound of coal compared with seven and a half for a good average boiler.

Froude was upset by the design of the suspension of the reciprocating parts of threshing machinery describing it as 'unprincipled' in not matching thickness to the stress at each part and concluding

> 'There is, to my mind, a professional immorality in ignoring this condition.'

The paper concludes with a lengthy discussion of mowing and reaping machines. Simple studies gave him an estimate of the work done in cutting with a scythe of 40 ft.lbs. per stroke, each cutting 6.25 sq ft at 27 strokes per minute giving a maximum work of 1080 ft.lbs./min. This is then compared with a mowing machine which, instead of 6.4 ft.lbs per square foot with the scythe used 41–126 ft.lbs. The cause of the power loss in the machine are then discussed at length, with numerical estimates. In this comparatively minor study, one can see the same principles of observation, measurement and analysis which Froude applied to his more famous investigations and one can also see the strength of his views on the 'immorality' of sloppy work.

3.5   Workshops

Froude clearly loved both designing and making mechanical devices. Henry Brunel, writing his obituary (ref 3.7), wrote:

> 'In his leisure Mr. Froude occupied himself a great deal with mechanical handiwork. His skill as a worker in material was great, and resulted from the educated knowledge of what should be aimed at, rather than from any particular excellence in that kind of aptitude which artisans acquire from practice. Even in ordinary work, he made use of well directed refinements of measurement for saving time. He was free from superstitious belief in the automatic accuracy of machine tools, and preferred to trust principally to gauges and surface plates, having a maxim that any error which could be detected could always, with proper care, be corrected. At the same time he did not neglect to employ all the advantages that good tools could afford. His lathes were kept in perfect order; there was no slackness, nor, what he seemed still more to dislike, any unnecessary tightness. Nothing was suffered to remain wrong. It must not, however, be supposed

that he was a slave to nicety of work and of fit. He knew well, where and when this was important; and he was never content with a suggested cause for the defective working of any machinery, until by putting the matter into quantities, he had satisfied himself that the cause was not only right in kind, but was also sufficient to produce the observed effect. He would often caution others against the temptation, as he expressed it, "to overestimate tendencies." His experimental apparatus always exhibited excellently finished work where finished work was necessary, and sufficiently, though less finished work in other places.'

3.6   Clockwork Model Tests

Froude's interest in ships and their performance continued and was even further developed. Abell says (ref 3.8)

'He appears about 1850 to have made screw propeller experiments on the River Dart above Totnes Wear, and also on the lake at Keswick (Bassenthwaite) constructing with great skill out of tin plate and solder most delicate and accurate recording apparatus.'

There is a further reference to these model tests in a letter to Edward Reed (20th January 1869) in which he said

'Years ago, Mr. Spedding assisted me in some experiments on screw propulsion which I made in one of the Cumberland Lakes.'
(This refers to Mr. James Spedding, younger brother of Thomas who married Phillis Froude)

Froude never published a full paper on these tests but they are outlined in his remarks on Rankine's paper to the INA in 1865 (ref 3.9). His opening words form quite a good description of what is now called the boundary layer.

'I so fully and entirely agree with the principles of all that he (Rankine) has brought forward. The ship rubs her side against the water and pushes the water forwards, the motion which she impresses upon the particles tends to produce motion in the line of the vessel's direction'

He sees the need to put the propeller in this current to recover momentum.

Froude then expanded his verbal remarks in an annex. There are several interesting passages; he introduces the word "wake" in its modern, technical sense and also uses the phrase, talking of the propeller, 'augments her resistance', foreshadowing his later work described in Chapter 13. He continues with a more detailed account of his clockwork experiments.

'Some years ago I tried a series of experiments bearing on these views, with a model 3 feet long to which I fitted a screw propeller driven by a carefully made piece of clockwork, the spring of which operated with a definite and tested force. The apparatus was so arranged that the propeller could be made either to occupy its usual position in the deadwood, or could be removed sternward, clear of the hull, to any distance short of about three-fourths of the extreme breadth of the model. A delicate floating logline, as it ran out, while the model was running, gave motion to a "paper cylinder", mounted amidships, on which a pencil, chronometrically driven, marked time while a separate arrangement recorded on it the number of revolutions of the propeller. The performance of the self recording apparatus was in every instance verified by a comparison with the actual length of logline run off, and the actual time occupied by the running. It will be understood, that, with this arrangement, as the same portion of the mainspring was used in every experiment, the propelling force was constant at all velocities – a circumstance which helped to simplify the results.'

Froude then 'lightly sketched the character of the results' summarised here. Moving the propeller aft from its normal position in the deadwood produced a very large increase in both speed and distance run for the same number of revolutions. Each movement aft increased speed though the rate of increase fell off, becoming negligible in the aftermost position. Since the propelling "force" was the same in every experiment, an increase in speed proved that there was a reduction in resistance thus showing that a propeller close to the stern develops an 'adventitious' resistance. Initial movement aft of the propeller produced a considerable increase in distance run per revolution (decrease in apparent slip) but a maximum was soon reached and further movement of the propeller reduced the distance run even though speed was marginally increased.

Froude then gives a rather confused description of the wake.

'On the whole, comparing the results obtained when the propeller occupied its normal position within the deadwood, with that obtained in its position of maximum advantage, the speed was increased from 0.98 knots to 1.4 knots, and the apparent slip was reduced from about 11% to 4%. This increase was equivalent, obviously, to more than would have been obtained by doubling the horse power of the model.'

With the propeller as far aft as it would go, Froude tried a dummy stern post abaft and then forward of the propeller. Performance was worse with the post ahead of the propeller. He suggested a full scale trial to confirm these tests.

Brunel and Lloyd (The Chief Engineer of the Admiralty) had somewhat similar ideas during the design and trials of RATTLER, the first RN screw warship, and it may be that Froude was aware of their work and thinking and that this inspired the tests at Keswick (ref 3.10). Though there is no direct evidence, it fits well. The account from a naval friend of dead water behind a screw battleship, described in Chapter 13, may also have persuaded Froude to undertake this work.

3.7 Early correspondence with J. H. Newman

During Froude's early years as an engineer he had little time for letter writing and Newman, too, was busy. Later letters suggest that they met from time to time to discuss their beliefs. Catherine had met Newman during his visits to Dartington and began to correspond with him, often passing on William's views with her own. Newman's 'Tract 90', published in 1840 was intended as a legalistic justification of the 39 Articles of the Anglican church but was seen by many as Newman's first step to the Roman Catholic Church (ref 3.11).

On 2nd April 1844, Newman wrote to Froude

'...on a subject which I have long wished to write to you about'; his thoughts and doubts. Newman was worried over the propriety of inflicting these doubts on others – 'It does seem unpardonable in a person like myself, who sits at home and speculates, to be thrusting his doubts on sincere and single minded persons like you and your wife.'

This introductory letter was followed by further long letters on each of the following three days and others at only slightly longer intervals.

The later letters were addressed to Mrs Froude, probably because William might be away from home. The replies were mainly from Catherine and showed clearly that she fully understood Newman's views and had a keen mind of her own, going far to disprove William's later claim that his wife was swayed by emotion rather than reason in religious matters. These letters mainly concerned the legalistic basis for the Anglican claim to be part of the 'True' church, on the nature of God and the monophysite controversy. Archdeacon Froude kept in touch with this correspondence with increasing disagreement.

In 1841 Newman leased an old stables at Littlemore, near Oxford, moving there in 1842. Meriol Trevor (ref 3.12) says

'Littlemore was held by Newman in joint tenancy with Wood, Church, Bloxham and William Froude. It was to be his retreat whilst he reconsidered his belief.'

William visited Newman at Littlemore in September 1845, only a few days before the latter was received into the Roman Catholic Church on 8th September, 1845. Most of Newman's Anglican friends, including Archdeacon Froude, severed all contact with him but William and Catherine remained close friends. Many years later (27th Nov 1868) Newman writing to Froude over the death of Sir John Harding, a mutual friend, said

'He was nearly the only person who was kind to me on my conversion (you were another).'

Newman took this correspondence with the Froudes very seriously as it was to them he had opened his heart and these letters were to influence his 'Apologia'.

3.8 The Achilli Trial

A steady correspondence on faith continued during the rest of the forties and early fifties, covered in the next chapter, only interrupted briefly in 1845 when a long letter from William, addressed to Newman in Rome went astray – it was returned to William by the Post Office in 1854. William thought he had offended Newman but by the beginning of 1847 their correspondence was back to normal.

There was a strong anti-Catholic feeling in much of English society, at all levels, with distorted memories of 16th century persecution fanned by false accounts of 19th century Catholic practices. One of the most prominent speakers against the Catholic Church was Achilli, a former priest, who left the Church, not because of doctrinal differences but for sexual misconduct. Newman attacked Achilli in a public lecture and was brought to trial for libel, then a criminal offence. In 1852 Newman's charges were found to be unsubstantiated and he was fined £100 with costs variously estimated at between £6000 and £13000, a very large sum in those days.

At the time William was in the Mediterranean, a visit which led initially to a fairly light hearted correspondence between Catherine, left at home with their five young children, and Newman over William's shocked response to the sale of indulgences. William was able to give some slight help to Newman in contacting witnesses to Achilli's character in the Mediterranean and was later to contribute £20 to Newman's costs.

The trial and verdict appear to have been most unfair, dominated by the prejudiced views of both judge and jury; even the Archdeacon told Catherine that

> 'he was certain that from the evidence, it was impossible that one man in the court could have a doubt of Achilli's guilt..'

Catherine continued 'that the Archdeacon seemed extremely disappointed at the verdict, – and was out of spirits all the day', perhaps showing that he was not too narrow minded. Her own view was that the trial afforded 'entire corroboration which the trial afforded of Newman's teaching that the intense prejudice against Catholicism has been more than a match for the natural love of justice in Englishmen.'

In March 1853 Froude wrote to Newman about anti Catholic prejudice:

> 'Perhaps much irritation is felt towards you by the bulk of the Anglican body; yet of this, a very large portion is due I am sure to that peculiar sensitiveness which is induced by a difficult or uncomfortable position. They are on slippery ground ... perhaps for the first time ... and you have shouted to them: and pushed them in the direction of least equilibrium; perhaps made them sensible that there was something awkward, ungainly and absurd

Chapter 3: Early Retirement

in their position... and for the most part I do not think the irritation of the better sort of Anglicans is of a different description ...'

3.9 Catherine's Conversion

During the 1840s and 50s, Newman continued to write to Catherine, pouring out his doubts and beliefs. Earlier, before her marriage, he had tried to enlist her active help in the Oxford movement and, since he knew that in 1839 she had written a highly religious novel which he thought showed some merit, he tried to persuade her to write children's' stories with a moral but she never contributed to the literary propaganda of the group.

During the early 50s, his letters were more clearly aimed at her conversion to the Roman Catholic faith and, indeed to demonstrate that her own views were already very close to that doctrine. William was unhappy over this turn in the correspondence and wrote: (3rd April 1854)

'But I must say something in reply to your kind letter about Kate. ..... I fully believe that as far as reasonable or reasoning conviction goes, her judgement is against Catholicism – as far as feeling goes it is in favour – the feeling being partly what might be called fascination occasioned by the magnitude and endurance of the system, and what appears to her to be the adaptions of this ceremonial to her own peculiar turn of mind and partly her own love and admiration for the Catholics she has known which goes entirely beyond that which she feels for any other persons whatever.'

'You will readily understand I am sure that it is with no cheerful feelings that I contemplate as almost a probable result, a change which, though it could not impair affection, would in its very nature make an end of that full community of thought and judgement in which affection has had such scope.'

From her letters it is clear that Catherine was an intelligent and strong minded lady and it may be that William was blinded by his own prejudice against the Roman faith, failing to realise that his wife's views also had a strong intellectual base, and that Newman valued her intellect as well as her sympathy. However, she wrote to Newman saying that she would not write to him again. Newman's response, to William, on 10th April, concluded:

'Let me add, that Catholics hold it would be wrong in any one becoming a Catholic without his judgement being convinced. .... Your dear wife has said she would not write to me again – and I assure you, my dearest William, I shall not write to her – but you can't hinder me (nor wish to hinder me) praying, whatever the prayers are worth.'

Within a month Catherine and Newman had resumed their correspondence, though the latter always reminded Catherine to show the letters to her husband. Within two years, Catherine was committed but there were some considerable practical difficulties.

Archdeacon Froude was still very strongly opposed to the Roman Church and Catherine sought permission, through Newman, for her to continue attending Anglican family prayers. This request was refused but on 19th March 1857 she was admitted to the Roman Catholic Church. She wrote to Newman,

> 'Convent of Mercy – Bristol   March 19, 1857
>
> My Dear Sir,
>
> I know that you will be glad to hear that I was received into the Catholic Church this morning. It is strange to think that you are the only person whom I now venture to tell of the great blessing which God has given me – not even my dearest William. ..'

In his brief and happy reply, Newman concluded, ' – and you must gain your husband by your prayers...'

William offered to leave Dartington and forego his considerable inheritance in favour of his youngest brother, Anthony. However, Archdeacon Froude saw Anthony as an atheist and less worthy than William who, with Catherine, stayed at the Parsonage, eventually receiving £35,000 under his father's will.

It is interesting to note that Catherine's conversion occurred at the same time as William's return to active professional work, perhaps as a relief from domestic differences. In 1857 he accepted two commissions from Isambard Brunel, one on rolling, dealt with in the next chapter, and the other on launching problems of the GREAT EASTERN.

3.10   Launching of SS GREAT EASTERN

The friction of the iron sliding launchways had been studied using an experimental cradle, loaded to represent the full scale one. Froude fitted 'self recording apparatus' consisting of a powerful pendulum, of very short period of vibration, formed from a piece of double headed rail, lengthwise, suspended on cords. A paint brush attached to the pendulum marked every quarter second on a tape attached to the moving cradle thus furnishing a record of motion, from which the exact amount of the retarding force of friction could be deduced for every moment of motion (ref 3.13)

Froude's experiments showed conclusively, that friction was not, as stated in the text books, independent of velocity, but became much less as velocity increased. The same recording apparatus was applied to the movement of the ship while being launched and similar phenomena were observed. During

Chapter 3: Early Retirement

Figure 3.1 – I. K. Brunel, one of the three men who influenced Froude

discussion of the launch of the GREAT EASTERN Froude pointed out how the fact of friction varying with the speed of the surface explained the analogous problem, that as soon as the action of a railway brake block reduced the speed of a wheel below that of the speed of the train, skidding ensued, and there was no alternative but to ease the brake until the wheel turned freely and re-apply the brake judiciously.

Henry Brunel wrote later (ref 3.14) that Scott Russell said, and one of Froude's letters to Bell seemed to support him, that I. K. Brunel had overlooked the fact that, as the ship moved down the ways, the cradles would wipe away the lubricant. In fact Froude's experiments showed that even at moderate speed the effect of lubrication was immaterial.

The work described in this chapter is varied and interesting and was carried out with Froude's usual care and penetrating analysis but it does not seem very much in total for a brilliant man in the prime of life.

References

3.1 See Annex 1.1.

3.2 Eliza's nickname was spelt 'Izy' by her father and, usually, 'Isy' by Henry Brunel (Sometimes Issy). I will follow her father.

3.3 W. Froude. Remarks on Mechanical Power and Description of a New Dynamometer. Bath and West of England Agricultural Journal, 1857. Papers, p.18.

3.4 W. Froude. On a New Dynamometer and Friction Brake. Trans. I. Mech. E. 1858. Papers, p.28.

3.5 W. Froude. Report on the Exhibition of Implements at Truro. Bath and West of England Agricultural Journal, 1862.

3.6 W. Froude. Report on the Exhibition of Implements at Falmouth. Bath and West of England Agricultural Journal, Vol. XVI, 1869.

3.7 Based on the Obituary by Henry Brunel, Minutes of Proceedings, Vol. LX ICE, 1880. (p.395).

3.8 Sir Westcott S. Abell. William Froude. Trans. Devonshire Association, 1933.

3.9 W. Froude. Remarks on: Prof M. Rankine. The Mechanical Principles of the Action of Propellers. Trans. INA., 1865. (Collected works).

3.10 D. K. Brown. Before the Ironclad. Conway Maritime Press, London, 1990.

3.11 The letters referred to by date between Newman and Froude are held in the Oratory, Edgbaston and are reproduced by permission.

3.12 Meriol Trevor. Pillar of the Cloud. Macmillan, London, 1967.

3.13 As 3.7.

3.14 Henry Brunel's Journal. Brunel Collection, Bristol University Library.

Chapter 4
Rolling 1857–1870

Part I – Development of the Theory

4.1 Introduction

Brunel was concerned that the GREAT EASTERN, so much bigger than any previous ship and with a different disposition of weights, might behave in an unexpected manner in a seaway. The second of the two tasks which he placed with Froude in 1857 was a theoretical study of rolling.

Rolling does not seem to have been seen as a matter of great concern either to seamen or naval architects prior to the early 19th century. Hull forms and loading were generally similar so that no one class of ship appeared greatly superior or inferior to others. The sails were quite effective in reducing roll as the rolling motion itself changed the velocity of the air over the sails in such a way as to oppose the roll (ref 4.1). By the mid 1830s rolling had become a matter of angry debate as the new Surveyor, Symonds, adopted proportions and forms which led to heavy and jerky rolling. Overall, the view seems to have been that rolling was a fact of life at sea and there was little point in studying it since nothing could be done.

The only serious works on the theory of rolling had been written over a century earlier. Several writers had considered the problem of a ship rolling in still water, without significant resistance to the motion, but only two men, Bernouilli and Don George Juan, had tackled the more difficult problem of rolling in waves. Both these men seem to have had some understanding of the physical nature of rolling but found the mathematics intractable and the simplifications they introduced led to incorrect conclusions. It is interesting that Froude in his letter to Brunel of 16th August 1857 implies that he was unaware of any previous works on rolling, a lack which he confirmed in discussion with Dr. Wooley in 1862.

4.2 Great Eastern

In parallel with Froude's theoretical study of rolling, Brunel asked William Bell, a brilliant member of his own staff (ref 4.2), to carry out tests using a carefully balanced model of the GREAT EASTERN. Bell had first to calculate the position of the ship's centre of gravity and then the distribution of weights about that centre (Moment of Inertia, defined in Annex 4.1). Neither calculation was difficult in principle but both were so lengthy that they had not previously been attempted; indeed,

Barnaby, a senior constructor and later Director of Naval Construction, said in 1865 that to calculate the position of the centre of gravity was so laborious as to be virtually impossible (ref 4.3). In discussion, Froude pointed out that Bell had carried out this calculation for the GREAT EASTERN. Froude had assisted Bell considerably in this work as shown in his letter to Brunel of 8th May 1857

> 'I will very gladly put all the data for the calculation of the great ship's moment of inertia (I hope she will be of more moment than of inertia) into Bell's hand and will co-operate with him.' (Froude's underlining)

Bell kept closely in touch with Froude so that their lines of inquiry soon merged, Froude writing to him in June 1857 to draw attention to an error in Bell's calculations, all too likely in such a long and tedious task. It is interesting that the calculations relate to the launch displacement of 11,600 tons suggesting that it was intended to carry out a full scale check on the position of the centre of gravity by means of an inclining experiment (ref 4.4) soon after launch. These preliminaries were complete by August 1857 when Froude wrote:

> 'Bell has got his tank and his model into something like working order (the model is very nice indeed) and with reference to a great many of the calculations which the case admits of we can by help of it verify results with great facility – we had a few preliminary trials a day or two since and I dare say a great many questions will be opened up in one's mind by observing experimental results.' (ref 4.5)

The model is believed to have been to a scale of 1/120 making it just under six feet long. The whereabouts of Bell's tank is unknown but Froude had a tank in the grounds at Dartington at this time and it may be that he is referring to this one.

4.3   Observation

Froude was just beginning his own investigation into the basic character of rolling in waves. Three letters to Brunel in August 1857 tell, with rising excitement, of his progress, based on observations of a tiny float.

> 'I hope shortly to be able to grind out in something like a tangible and intelligible shape the quest of the movement of a ship on the side of a moving sea.' (9th August)

which suggests his ideas were incubating. A long letter on 16th of the same month brings out the first key discovery, (underlining is Froude's)

> 'I cannot forbear writing you a line to communicate about what I consider to be a discovery of mine and a highly valuable one viz that at the surface of a wave gravity (or with the resultant of gravity and the other forces by which a body floats on the surface

Chapter 4: Rolling 1857–1870 : Part I – Development of the Theory

of the wave will be achieved) is at right angles to the surface of the waves so that if you are on a raft, a plumb line will hang at right angles to the wave surface and not in a direction perpendicular to the earth's surface. I was led to this conclusion by considerations of general dynamics but I think I may truly say I have already tested it by experiment for I have taken a ring of cork, stuck together and hung on a copper wire, bent to a circle of 2 inch diameter; (fig 4.1) erected three slight sheers on this, with a plumb line hung from the apex to the water line – and placed them in a basin of water, I find that if the water be set into steady oscillation, reaching considerable angles of obliquity, the bob nevertheless remains quite steady in the centre of the rig – as sensibly steady as if being tested on a fixed table at least so long as the oscillations are large and steady compared with the size of the floating ring for the bob becomes unsteady (as indeed it ought to do) when minor waves are introduced such as to affect one edge of the ring much before the other is affected.'

This was followed by another long letter on 21st August which began with a rare example of Froude's somewhat laboured humour.

'I am taking off on a cruise in a yacht for a week and perhaps shall have to study rolling and pitching in a manner that will impress itself on my recollection. But before I compose my understanding by the action of my stomach, ...'

He continued by saying that he had verified the conclusions of the previous letter by tests in waves generated in the river by a small boat and by a paddle steamer. More important, he demonstrated that when a wave acts bodily on a ship, the rolling effect will be 'greater in proportion as her stability is greater.' He confirmed this with tests on a cork float, with a rod through it, on which a weight could be moved up and down to alter its stability showing that 'if it (the weight) is raised till the stability is almost gone, it floats almost undisturbed, ie the greater the force acting to restore a keeled ship to the upright, the worse will be the rolling.'

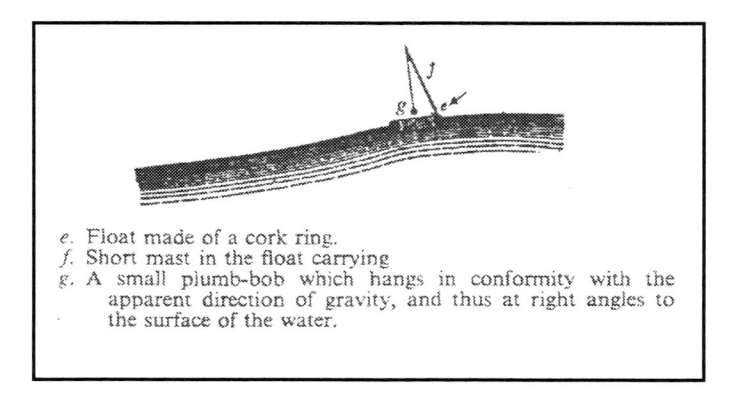

Figure 4.1 – A sketch by Froude showing how the pendulum on his tiny float hung perpendicular to the wave surface ...

e. Float made of a cork ring.
f. Short mast in the float carrying
g. A small plumb-bob which hangs in conformity with the apparent direction of gravity, and thus at right angles to the surface of the water.

The Way of a Ship in the Midst of the Sea

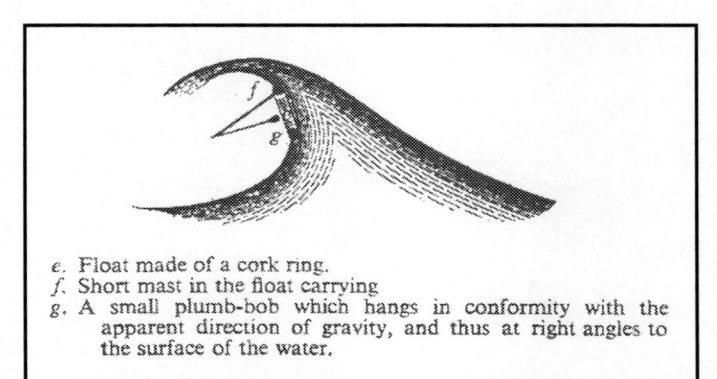

e. Float made of a cork ring.
f. Short mast in the float carrying
g. A small plumb-bob which hangs in conformity with the apparent direction of gravity, and thus at right angles to the surface of the water.

Figure 4.2 ... even under a breaking wave

These illustrations are reproduced from Froude's original paper From the dates, this test may be related to a story told by Harper (ref 4.6) about an experiment in an estuary near Salcombe in 1856 when he was accompanied by two naval officers in uniform. He says that the locals had long believed that Froude was mad and when they saw him accompanied by two men in uniform they assumed that he had been certified insane and that the officers were his keepers.

These simple experiments well illustrate Froude's approach to a novel problem. Observation of the behaviour of his tiny float in waves led his analytical mind to perceive many of the important aspects of rolling behaviour, though it should be remembered that the experiments themselves were initiated following his consideration of the dynamics of rolling. The next step would be a mathematical analysis of these observations which would lead to a generally applicable theory of rolling. He wrote to Brunel on 4th March 1858.

> 'I will only add that I am in strong hopes that, with the help of Bell whose mathematical tools are in some respects less rusty than mine, I shall be able really to get to the bottom of the "wave question" in which one or two real steps have been made between us lately.'

The work was delayed as in late 1857 and the first half of 1858 Brunel and his team were fully occupied with the problems of launching the GREAT EASTERN. Then, on 15th September 1859, Isambard Kingdom Brunel died.

4.4   A Theory of Rolling

Froude and Bell, who from 1858–61 was resident engineer on the Torquay to Dartmouth railway, developed the mathematical treatment of rolling, based on Froude's observations of his little float. This work was presented on 1 March 1861 as an "extemporary address", 'On the Rolling of Ships', (ref 4.7) to the Institution of Naval Architects (INA), founded the previous year. This was revised and extended before publication in the Transactions, a demanding task which took him some time; several letters from Henry Brunel (Isambard's second son) pressing him to progress the work more quickly.

It was Froude's first published paper dealing with hydrodynamics and not easy to describe in simple language as it is highly mathematical. Froude's treatment is, to use a word beloved by mathematicians, 'elegant' and readers miss a great deal in this simplified account which will concentrate on Froude's approach to the problem, his conclusions and their value.

Like the earlier workers, he had to adopt a simplified picture of a ship moving in waves. He assumed that the ship was stationary, beam on to regular waves, all of the same length from crest to crest, and with the same period, and resistance to rolling was ignored. It was also assumed that the moment of the force trying to bring the ship upright was proportional to the roll angle and, in consequence, the roll period was the same for all roll angles, known as "isochronous" rolling.

These simplifications are important; a comprehensive mathematical treatment was and remains impossible and while earlier workers had simplified their treatment in such a way that the results were erroneous, Froude's approach led to conclusions which were generally sound. Even his simplified picture of rolling was able to explain most aspects of ship rolling and to quantify some of them and, because it was basically correct, he was later able both to extend his mathematical treatment and also to develop empirical methods for those aspects which could not be treated mathematically.

His observations had shown him that a small float behaved as though it was a particle of water forming the wave and he reasoned that, compared with an ocean wave, even a ship was small and could be treated as though it was a particle of water. One is reminded of the fisherman's prayer

'O Lord, thy sea is so big and my ship is so small.'

The conclusions drawn by Froude are lengthy as he was at great pains to point out the conditions under which they were valid in the light of the approximations involved.

The rolling of a ship has much in common with the behaviour of a pendulum and this analogy will be used, as far as is possible, to explain the results of Froude's work. The natural period in which a pendulum swings depends on its length, a long pendulum swinging more slowly. The short pendulum with a rapid swing is said to be 'stiff'; stiffness depending on the inverse of the length. The roll period of a ship is discussed in Annex 4.1 where it is shown that the period depends on the inverse of the metacentric height, ie a ship with a small metacentric height (low stability) will have a long roll period. (ref 4.8)

The height of the centre of gravity, G, depends on the arrangement of weights, both fixed and variable (such as cargo or fuel), while that of the metacentre, M, depends on the underwater shape. The vertical separation of G and M, (GM), is known as the metacentric height. Froude's first conclusion was that, if the separation of G and M was the same regardless of their absolute height, then the ship would roll in the same way in identical waves. (Annex 4.1)

Most people will have played with a pendulum or a child's swing and discovered that a series of gentle pushes, applied regularly, at the same point in the motion, will cause the swing of the pendulum to increase very rapidly. This coincidence of the application of the disturbing force with the natural period of the pendulum is known as resonance and one of Froude's early discoveries was that rolling is worst when the period of the ship is in resonance with that of wave encounter and under such conditions the roll angle can increase alarmingly. This had not previously been appreciated though it is implicit in Bernouilli's work.

Froude's third conclusion was that ships with a long natural period, associated with a small metacentric height, will suffer least from rolling and he pointed out several reasons for this. Long period waves take a considerable time to build up and are less common than waves with moderate periods and as a gale dies down, long period waves, which travel faster, go away more quickly. Long period waves are also long from crest to crest and hence are usually less steep than shorter waves, disturbing the ship less.

Perhaps more important, a stiff ship, with a short period, will build up roll more quickly when regularly disturbed, as may be found by experimenting with different lengths of pendulum. Because of the effects of resistance to roll, very stiff ships, such as those designed when Symonds was Surveyor, though they will usually roll to only slightly greater angles, will roll much more quickly, with greater accelerations. These accelerations will increase the loads on the masts etc and make accurate gunnery difficult. The balance organs in the human ear are confused by accelerations and inform the brain that the roll angle is considerably greater than the real angle whilst the readings of a pendulum can be deceptive for similar reasons.

The compromise between sufficient stability (GM) to prevent excessive heel under wind loads, stability at large angles to resist capsize while avoiding severe rolling is never easy. The naval architect uses the word 'stability' in a specific sense; the greater the metacentric height, the greater the stability. This leads to the apparent paradox that a 'stable' ship will roll more severely than one which is less stable but this paradox is merely a matter of semantics, depending on the meaning attached to stability. In Froude's day and until the advent of the computer it

Chapter 4: Rolling 1857–1870 : Part I – Development of the Theory

was more difficult to balance the conflict between safety and easy rolling as the relevant calculations were long and complicated while their results could only be judged against long experience in service which was not available in 1860.

There are, in fact, two ways of giving a ship a long natural period. The first is to move weights as far as possible from the roll axis, increasing the moment of inertia. The heavy armour on the side of ironclads did just this and served to moderate their roll, not enhance it as was widely believed at the time. In unarmoured ships the scope for changing the moment of inertia for roll is small.

The alternative approach, already discussed, was to reduce stability (metacentric height). Froude points out that this can be achieved by raising weights and hence the height of the centre of gravity 'consistent with her performance of her regular duties'. This was seen as the most controversial part of his theory and Froude was accused of advocating ships with poor stability. He was right in principle; a small metacentric height will generally reduce the severity of rolling. The phrase quoted above shows that he appreciated that there were practical limits to this approach though there are indications that he did not, at first, fully recognise the severity of those limits. It is certain that this aspect of his theory was misunderstood by many designers and as a result there were a considerable number of ships, particularly liners, built with inadequate stability in the hope of reducing the severity of rolling.

Sir Edward Reed, Chief Constructor of the Navy wrote (ref 4.9)

> 'There is no feature in the performances of our ironclads which has been so much misrepresented and misunderstood as that of their rolling at sea. From the reports and criticisms of some persons, it would appear that these ships roll to an extent which is most excessive in comparison with wooded line-of-battle ships and frigates;..'

He goes on to say that, in fact, several are remarkable for their steadiness and gives credit to

> 'The labours of Mr. Froude who has within the last few years taken the lead in this matter...'

Reed tried to apply Froude's work before it was fully developed and the AUDACIOUS class, which he intended should have a fairly small metacentric height, completed with too little and required 360 tons of cement ballast. However, they were still notable for their steadiness at sea, making them good gun platforms.

Froude confirmed some of these results by tests on three geometrically shaped bodies in a trough fitted with a wave maker which was driven by a crank off a hand turned flywheel kept in pace with the swings of an adjustable pendulum. These experiments came to an end when he moved house in 1859, from Dartington to Paignton, on the death of his father.

The 1861 paper also gave some numerical results showing that a small ship, with a roll period much less than that of the waves which she encountered would roll away from the wave with an angle to the vertical only a little greater than the wave slope. A bigger ship, with a longer roll period, encountering waves with a period only slightly less, would roll towards the wave with a very great angle. The neglect of resistance to rolling in this early work of Froude's meant that he greatly overestimated the maximum angle of roll in this case. Even if the wave period was much less than that of the ship it would still roll into the wave. Examples like these justified his contention that rolling was worst in ships with too great a metacentric height. It is interesting that Froude had written to I. K. Brunel on 8th May 1857 commending the work of the Master of HMS DUKE OF WELLINGTON. Moriarty, for his trials of metacentric height and rolling, summarised in Annex 4.1.

4.5   Froude and Mathematics

A surprising aspect of this 1861 paper is that Froude was 48 years old when the mathematical work was started in 1858 and yet it is very rare for any one over about 30 to produce innovative mathematics particularly if they have not been continually involved in such work. Froude's letter to Isambard Brunel in March 1858 envisaged help from Bell and in writing the final version of the paper in 1862 this help is fully acknowledged as

> 'an able mathematician and an accurate calculator who I am bound to say, not only relieved me of all the laborious part of the detailed numerical calculations relating to the ship, but joined so heartily and effectually in the theoretical part of the investigation that there are very few parts which I am inclined to call specially my own.'

Though in a footnote, Froude made it clear that he had a mathematical mind of his own.

> 'To my friend, Mr. Bell,...., I am indebted for the following integration, the key to which he lighted on in a volume of professor Airy's. I have altered it from the form in which I received it, to one which, in some of its steps, appeared to myself somewhat clearer; I only hope he will not consider that I have attempted to secure it as my own by spoiling it." (ref 4.10)

Froude had earlier written to Brunel on 15th May 1857 (ref 4.11) on Bell's ability.

> 'Bell is very rapid and correct in the use of figures: much more so than I am myself and though a little too reliant on formulae perfectly "honest", a qualification which I agree with you is most important in such inquiries and as rare almost as it is important

Chapter 4: Rolling 1857–1870 : Part I – Development of the Theory

for were it more common nearly all the absurd theories which are current in the world would die of inanition – yet I suspect few people can have watched the working of their own minds in such questions without feeling how difficult it is to be honest.' and – 'Perhaps you will remind me of a proverb about the crock and the kettle if I say he is a little too much given to reliance on equations.'

An interesting reflection on Froude's own thinking.

The paper was by far the most mathematical to be presented to the INA in its first decade but it would seem from some of the footnotes that Froude not only provided the basis for it in his observations and original analysis of 1857 but was also a full partner in the mathematics. (See Chapter 15 for further discussion of Froude's mathematical ability)

Bell was obviously well content with Froude's acknowledgement of his assistance as he continued to help for several more years and remained a friend for life.

4.6 Debate

The discussion on Froude's paper was published in the 1862 Transactions of the INA (ref 4.12). Of those who had knowledge of rolling, Scott Russell, the designer of the GREAT EASTERN, accepted Froude's view on the effect of metacentric height though he maintained, very strongly, that ships' form dominated rolling behaviour. Canon Mosely, a mathematician who had made an important contribution to the theory of large angle stability, supported Froude on the importance of synchronism between wave and ship and mentioned experiments which had been carried out in Portsmouth Dockyard at his request.

Crossland, an Admiralty Constructor and an outstanding mathematician, made a very lengthy contribution which was almost a paper in itself. In reply Froude said:

'Mr. Crossland appears to me to have followed out, on an independent line of thought, and in a thoroughly common sense and sound manner, a series of steps by which the fundamental laws of dynamics may be applied directly to the elementary conditions of motions belonging to wave surfaces, and to have pursued them further, in a manner not less original and interesting, into the more intricate and difficult conditions of complete wave motion – arriving, not indeed at a complete solution of the question, but at many very instructive views concerning it.'

However, Froude thought that Crossland had 'not fully apprehended ... that my mode of dealing with the subject is avowedly approximate.' The well respected Rev. Joseph Wooley, another mathematician, seemed to accept very little.

4.7 Amplification

In a series of papers over the following few years, Froude tidied up some loose ends in his theory. The first contribution came in the discussion of a short paper by Scott Russell in which the author tried to distinguish between the effects of stability due to weight and that due to form (ref 4.13)

Froude's response ran to a massive 43 pages in which he showed that Scott Russell's examples were extreme cases, irrelevant to real ships. He re-iterated that stability must be adequate for safety but not excessive. He drew attention to the need to increase resistance to rolling such as the use of bilge keels – he called them 'bilge pieces' – and mentioned several experiments, including some in which he had confirmed the performance of Scott Russell's more extreme forms.

Perhaps the most interesting part of this contribution lies in a footnote. The footnotes to Froude's papers are usually well worth study as they show the breadth of his thinking and, as they are often more personal in style, give some idea as to how his thinking developed. In this discussion there is a footnote which sets out the conditions necessary for scaling model results to ship size, later known as Froude's Law of Comparison. This will be dealt with in Chapter 9.

Froude read two other short papers (ref 4.14) to the Institution in 1863, one on the significance of isochronism in terms of real ship forms and the other an interesting side line on waves at the interface between liquid layers of different density. In 1865 he attempted to estimate the maximum roll angle which would be reached by a ship in a given sea state. He hoped to show which ships were in danger from capsize but the basic assumptions, neglecting resistance to rolling, were too simplistic for the results to be meaningful though some useful guidelines were given.

This paper marks the end of the first phase of Froude's work on rolling; he had solved the mathematics of unresisted rolling in regular waves and, subject to a few minor changes, had demonstrated that his work was sound. His work now turns to the study of resistance to roll, mainly by model testing, and of confirming the results by measuring rolling at sea. It was also the occasion for a very nice compliment from Scott Russell who had chaired the 1863 meeting.

> 'I cannot allow the discussion to close without saying how highly indebted we are to the perseverance with which Mr. Froude has continued to attack difficulty after difficulty in this most difficult of subjects. I think it is very encouraging to find each time when we come together that something that was almost impossible last time has been done; that we live, in short, in a time when the impossibilities of yesterday become the achievements of today.'

Chapter 4: Rolling 1857–1870 : Part I – Development of the Theory

Froude and Scott Russell were not friends and differed considerably over professional matters but in public they were more than merely polite (ref 4.15) Emerson, in his well known biography of Scott Russell, indicates considerable hostility between the two men. This is based mainly on references in letters by Henry Brunel to Froude but Henry's language in writing of others was always highly coloured and it does not necessarily follow that Froude fully shared his views on Scott Russell.

References

4.1   William White. Manual of Naval Architecture. John Murray. London, 1900.

4.2   William Bell was born at Leith on 21st September 1818 and graduated brilliantly from Edinburgh in Natural Philosophy and Mathematics in 1839. After early experience as a railway engineer, he joined I. K. Brunel in 1846, initially on the Bristol Docks. By 1850 he was involved with original theoretical and experimental work on the design of the Chepstow bridge. Later, he was involved in further investigations on bridge design and on the strength of timber for the South Devon railway.

He seems to have acted as Brunel's research man, often being called in on a problem demanding unusual mathematical or experimental skill which is why he was involved with Froude on rolling. Bell wrote the chapters on bridges and docks for the Life of Isambard composed by his sons and contributed to William Froude's obituary. In later years, Bell made further notable contributions to bridge design as a consultant. He died on 20th January 1892. (Trans. ICE. 1892)

4.3   N. Barnaby. An Investigation of the Stability of HMS ACHILLES. Trans. INA., London, 1865. (Discussion)

4.4   See Annex 4.1.

4.5   Brunel collection. University Library, Bristol. It has been thought unnecessary to reference individually letters which are identified by date in the text.

4.6   Gordon Harrington Harper. Cardinal Newman and William Froude – a Correspondence. John Hopkins University, Baltimore, 1933.

4.7   W. Froude. On the Rolling of Ships. Trans. INA., 1862. Included in collected works but as perhaps his greatest work it deserves a special notice.

4.8   See Annex 4.1.

4.9   E. J. Reed. Our Ironclad Ships. John Murray, London, 1869.

4.10 As 4.7.

4.11 As 4.5. Report book IV, 1856–7.

4.12 As 4.7.

4.13 J. Scott Russell. The Rolling of Ships as Influenced by their Forms and by their disposition of weights and W. Froude. Remarks on Mr Scott Russell's Paper on Rolling. Trans. INA., London, 1863.

4.14 The Papers of William Froude 1810–1879 Institution of Naval Architects, London, 1955. This contains 37 of Froude's papers and, since copies are not uncommon, it has not been thought necessary to reference papers individually in most cases.

4.15 G. S. Emerson. John Scott Russell. John Murray, London, 1977.

Annex 4.1 – A Note on "Stability"

Small Angles of Heel

Initial Stability If a ship is heeled in still water to a small angle, about 10°, then the shape of the underwater body will change and the new line of action of the force of buoyancy, acting vertically through the centroid of the immersed volume, will cut the centre line at a point called the metacentre (M) (fig 4.A1). For these small angles of heel the metacentre can be treated as a fixed point. The vertical separation of M and the centre of gravity, G, is called the metacentric height and is used as a measure of the moment trying to bring the ship back from the heeled position to the upright. This approach is often called the metacentric theory and was well known by the late 18th century.

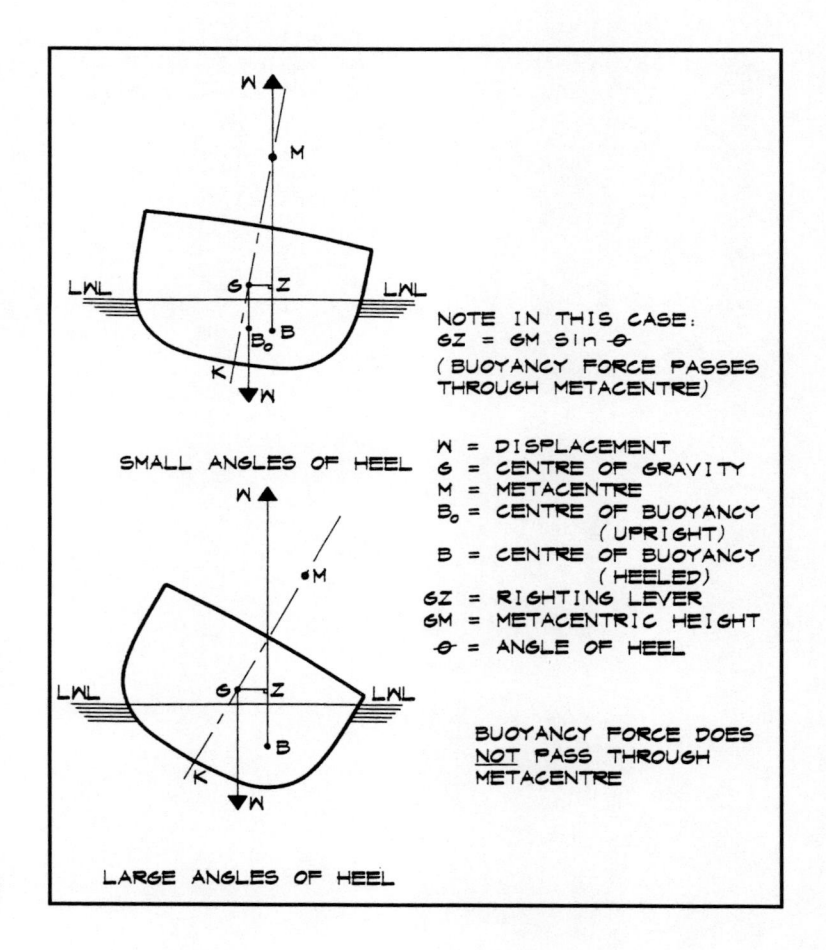

Figure 4.A1 Static stability relationships

Righting moment = Displacement x GM x Angle of Heel (Radians)

Chapter 4: Rolling 1857–1870 : Part I – Development of the Theory

This righting moment governs the 'power to carry sail' and the convention is used that the greater the metacentric height, the greater is the stability. The height of the metacentre depends to a very large extent on the beam and can be calculated from the geometry of the lines plan. Before computers, this was a fairly lengthy task but not difficult.

The Inclining Experiment The value of the metacentric height can be found by moving a known weight, W, a distance, d, across the deck. The angle of heel is measured with a pendulum and when the ship is still:

W x d = Displacement x GM x Angle of Heel (Radians)
giving GM and hence the position of the centre of gravity.

The principle of the inclining experiment was known and published in the late 18th century but was rarely used until after the loss of the CAPTAIN. (Chapter 5)

In a letter to I. K. Brunel of 8 May 1857 commending the work of H. A. Moriarty, the Master of HMS DUKE OF WELLINGTON, Froude describes the experiment carried out by Moriarty. He moved the guns weighing 150 tons through 4.5 feet, measuring 4.5 inches deflection on a 15 foot pendulum (1.5°).

> 'and though the angle is but small, yet the day was so fine and calm, and other circumstances were so favourable that I believe the experiment is nice enough to determine the cg pretty exactly.'

Moriarty also

> 'observed very carefully and repeatedly the natural period of the ships rolling in still water and compared this with the period when at sea and with the periodic recurrence of waves, when this was such as to give the maximum rolling. The period in still water, with the guns run out, was 14 seconds and with the guns housed, 12.5 – 13 seconds. On passage with lower and middle deck guns housed the period was about 13 seconds and the maximum rolling depends on the recurrence of waves being concurrent with this.'

The rolling was lessened and almost ceased whenever heavy seas recurred in quick succession. Unfortunately, the letter does not indicate whether Moriarty had carried out this work on his own initiative, as seems likely, or whether it had been suggested by Froude.

Rolling The natural period of roll of a ship depends primarily on the metacentric height (GM) and on the Moment of Inertia (I). The moment of inertia is calculated by multiplying the weight of each element of the ship by the square of its distance from the axis about which it rolls and adding them all together, a very tedious task which Bell carried out for the GREAT EASTERN. It is often convenient to use the radius of gyration (k) where I = Displacement x k^2

The period of roll, out to out and back (Froude used the time from out to out only) is given by

$$T = 2\pi \sqrt{\frac{k^2}{g.GM}}$$

For modern warships, a convenient approximation is

$$T = 0.78.B\sqrt{GM}$$

Stability at Large Angles of Heel The resistance of a ship to capsize depends on the righting moment at large angles of heel and on the angle at which the righting moment reaches its largest value before beginning to diminish with further heel. At these larger angles the shape of the underwater part of the of the ship changes radically and the line of action of the buoyancy no longer passes through a fixed point. (metacentre)

The mathematics of large angle stability had been set out by Attwood in the late 18th century but the work involved in the calculation of even one value was enormous. The calculation of a full curve of righting moments (fig 4.A2) was first made possible by the work of Barnes in the late 1860s. Such a calculation involves a considerable number of assumptions and

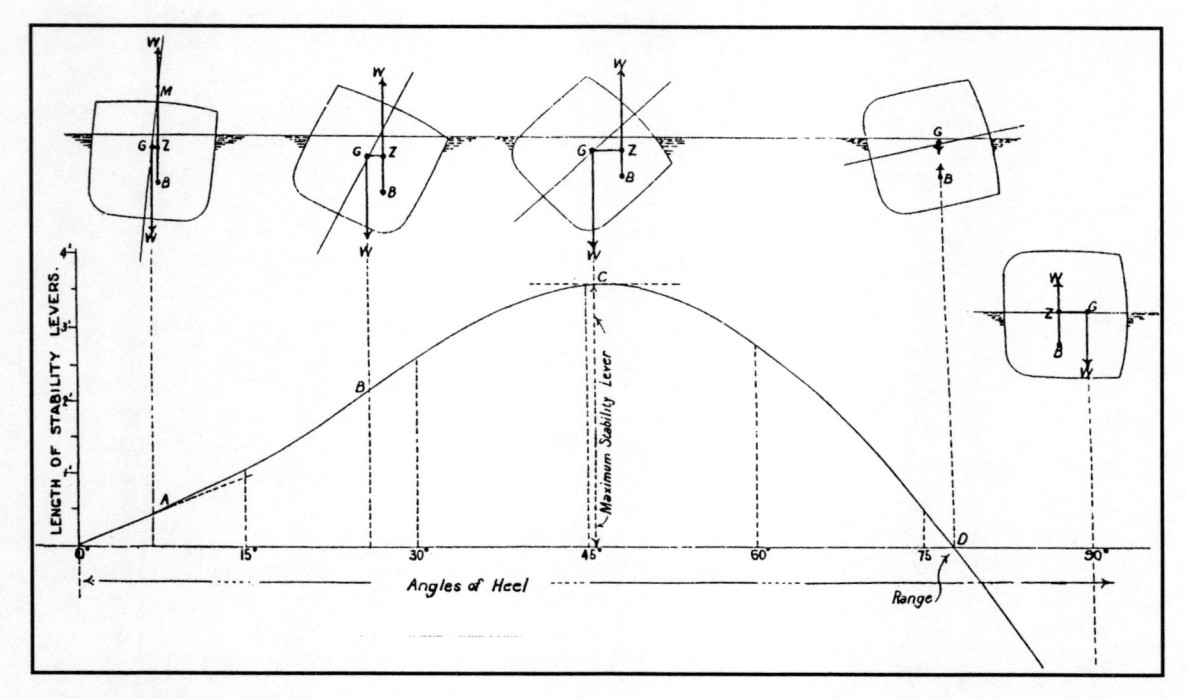

Figure 4.A2 – Curve of Stability
The small sketches show how the righting lever (GZ), the horizontal separation between the equal and opposite forces of weight and bouyancy, vary with angle of heel; increasing at first and, after a maximum, falling to zero. The curve below shows a continuous graph of GZ against heel.

Chapter 4: Rolling 1857–1870 : Part I – Development of the Theory

criteria of acceptability can only be based on considerable experience – indeed the current criteria are based on a tragic episode in the USN when three destroyers capsized in the typhoon of December 1944.

The distance between the lines of action of weight and buoyancy is the righting lever (GZ on fig 4.A2) and the righting moment is Displacement x GZ. For small angles of heel GZ = GM x Angle of Heel (Radians)

A graph of righting lever against angle of heel tells the designer a great deal about the safety of the ship. As the heel begins to increase, GZ and hence righting moment increases quite rapidly. When the deck edge is immersed, increasing heel does not help righting moment to the same extent and a maximum value of GZ is reached. Further heel leads to a reduction in righting moment which will diminish to zero and become negative.

A ship may pass beyond the maximum righting lever while rolling and recover. On the other hand, a ship exposed to a steady heeling force, as of the wind on sails, will be safe only up to the maximum righting moment. If the maximum is less than the heeling moment, the ship will capsize as did CAPTAIN. Righting moment and GZ at large angles are much more dependent on freeboard than on beam or metacentric height. A ship which is unlikely to capsize may also be said to have good 'stability' though the meaning is different from that used earlier.

It should also be noted that a ship which does not roll much is often said to be 'stable' though, as Froude showed, such usage is the opposite of the meaning used in this annex.

The Way of a Ship in the Midst of the Sea

Chapter 5
Rolling 1870 – 1875

Part II – Application and Trials

5.1   Wider Horizons and Problems

From about 1864, Froude was increasingly involved in work on the powering of ships, both for the British Association and on the tests of his 'Swan and Raven' models, (Chapter 9). From 1868 this merged into his negotiations with the Admiralty for the towing tank at Torquay.

There were also the personal problems discussed in Chapter 6 including his son Edmund's wish to become a Catholic priest which upset William greatly and also made him fear that he would lose Edmund's assistance. Also in 1864 Froude's daughter, Mary, died followed in 1865 by Henry Brunel's affair with 'Polly' which strained, but did not break, his friendship with the Froudes.

5.2   Loss of the CAPTAIN

The new battleship CAPTAIN capsized on 6th September 1870 with the loss of nearly all of her crew and arguments over stability, rolling and capsize ceased to be academic. The story of the design, building and tragedy of the CAPTAIN is long and complicated but, since it had important interactions with Froude's work, an outline must be given here (refs 5.1, 5.2).

Based on his experience during the Crimean War, Captain Cowper Coles saw the need for heavy guns to be mounted on a turntable – 'turret' – and Admiralty support led to tests of a prototype (built by Scott Russell) in HMS TRUSTY in September 1861. These showed that a turret gun could be fired more rapidly and more accurately than a broadside gun and that the turntable would not be jammed by the impact of the heaviest shot (ref 5.3).

As a result two coast defence vessels with turrets were ordered early in 1862 and, together with the results of the later battle between USS MONITOR and CSS VIRGINIA, (usually known by her former name, MERRIMAC), faith in the turret was enhanced. Coles pressed for a fully rigged ship, with a large number of turrets on the upper deck. This design was quite impractical and led to considerable ill feeling between Coles and the Constructor's Department and with the Chief Constructor of the Navy, Edward Reed, in particular. In 1865 the Admiralty set up a 'Committee of Naval Officers' to consider Coles' claims and they concluded that a seagoing turret ship should be built, with a moderate freeboard and good sea keeping. This ship, MONARCH, was designed by Reed and had a long and successful life. At this date, there were frequent changes in the political head of the Admiralty – First Lord – which may partially explain the muddled decision making: a list follows (ref 5.4).

Date Appointed	First Lord
June 1859	Duke of Somerset
July 1866	Sir John Pakington
March 1867	Henry L. Corry
December 1868	Hugh Childers
March 1871	George Goschen

Coles was not satisfied; he wanted a low freeboard so as to reduce the extent of side to be armoured and he also objected to MONARCH's forecastle, seen as essential for seaworthiness, but which prevented the turret from firing ahead. After heated discussion, the First Lord, then the Duke of Somerset, decided that Coles could select a shipbuilder from a list supplied by the Admiralty to develop his ideas into a design and, as a result, Laird's submitted designs in July 1866 with a forecastle but with very low freeboard.

Reed and the Controller, Sir Spencer Robinson, objected strongly to the design on the grounds of its low freeboard and fears that the weights had been under estimated. They were over ridden by the new First Lord, Pakington, who was determined to show that industry could outdo bureaucracy and CAPTAIN was built by Laird's. It is a long and complicated story (ref 5.5) from which only Spencer Robinson emerges with total credit. The other players all 'improved' their original opinions after the ship was lost though, to be fair it is likely that Reed's engineering intuition told him that the freeboard was inadequate, a fear which could only be explained after Barnes had developed his method for calculating large angle stability. Stability is discussed in Annex 4.1; in brief, the force tending to bring a ship back to the upright depends on beam for small angles of heel but on freeboard at larger angles, something not fully appreciated in the 1860s.

It should be noted that Reed was not opposed to turrets per se and designed a number of successful turret ships himself, such as DEVASTATION, discussed a little later. However, he did not see the upper deck of a full rigged ship with its maze of ropes as an appropriate site for a turret. On completion, CAPTAIN was grossly overweight, mainly because Laird's original estimates were greatly in error but their weight control was also poor and hence the inadequate design freeboard of 8 feet was reduced to 6ft 7ins.

Initial reports from sea were very favourable but an inclining

Chapter 5: Rolling 1870–1875 : Part II – Application and Trials

experiment was carried out at the end of her second voyage to check her stability which was already suspect. This work was extended by Barnes and Barnaby of Reed's staff to consider her stability at large angles of heel, only the second time this lengthy calculation had been made. This calculation was, and remains comparative rather than absolute and with no previous ship safe or unsafe for comparison, it was not possible to be sure if CAPTAIN was unsafe until her capsize ended discussion. Her maximum righting moment came at a heel of about 18 deg and she would not recover if blown past this angle by a sustained force. During her last day CAPTAIN had been sailing with a heel of 17–18 deg, close to the limit of safety, and round about midnight on 6th September 1870 a gust blew her over with the loss of 473 lives, including that of Coles.

Reed had already resigned on 9th July 1870 in disgust over the methods of yet another First Lord, Childers. Spencer Robinson was forced by Childers to follow suit, the two men who had warned of the danger. One is reminded of a current statement of the stages of a project in which disillusionment is followed by

'The search for the Guilty and Punishment of the innocent.'

Froude's own views on the loss of the CAPTAIN were rather simplistic and given in evidence to the Royal Commission of 1872 in the context of the cost of scientific research read:

'I may say that the value of such services when rightly applied, enormously outruns any possible expenditure that they might involve. The cost of that single disaster, the loss of the CAPTAIN, exceeds immeasurably, any kind of expenditure that could be devoted to the improvement of science as applied to naval architecture...........As it was, Mr. Reed's strongly urged objections to the ship, though based on scientific grounds, were overborne by ideas which relied simply on seamanship and traditionary knowledge.'

In considering whether a Scientific Council would have been effective in supporting Reed, Froude said 'Captain Coles had many friends, he was a very able seaman, and he had the facility of inducing people to believe in him, and it was then believed that his practical knowledge as a seaman enabled him to judge of questions which were entirely beyond the cognizance of mere practical seamanship.'

5.3 The Committee on Designs 1871

After this disaster the Admiralty set up a 'Committee on Designs for Ships of War'. It was to consider the safety of existing warships, and those currently building, for the true cause of CAPTAIN's loss was not understood by many people and it was also to make recommendations on the style of future ships. The Committee comprised some of the most outstanding engineers, scientists and naval officers of the day. It started taking evidence in March 1871 and produced a long, detailed and generally sound report in July of that year.

Froude was a member of this Committee and also gave scientific evidence, demonstrating official recognition of his talents (also marked in 1870 by election as Fellow of the Royal Society) and providing the opportunity for further work as he was asked by the Committee to carry out rolling tests on a model of the DEVASTATION, a new, low freeboard, turret ship but totally different from CAPTAIN in that she did not carry sails.

Froude's main work for the Committee was the extensive set of experiments on the DEVASTATION but, before describing them, mention must be made of some lesser tasks. He carried out rolling trials on the new coastal defence ship, GLATTON, in December 1871. These trials, at Chatham, seem to have been the first full scale rolling trials. The ship was rolled by men running across the deck until angles of from 7 – 12 degrees were built up. Henry Brunel was Froude's assistant and wrote in December 1871 that he would have liked to work without charge but 'some coin would be useful' to pay his builders.

The Committee had already realised that the stability at large angles was inadequate in the somewhat similar CYCLOPS class and recommended fitting watertight superstructures to increase freeboard. It was noted that their rolling would be excessive in waves with a period longer than 10.5 seconds, and though the record is unclear, this can only have been suggested by Froude.

5.4 Model Rolling Tests of DEVASTATION

The DEVASTATION was the first large ship built for the Royal Navy without sails (fig 5.1). She was of a style developed by Reed in coast defence ships with a low freeboard hull, heavily

Figure 5.1 – HMS DEVASTATION used for rolling trials. It is very hard to see how 400 men could run backwards and forwards across her obstructed decks.

Chapter 5: Rolling 1870–1875 : Part II – Application and Trials

armoured on the sides, and an armoured breastwork above the hull carrying the turrets, access and ventilation openings, together with the bridge and funnels. The Committee considered that her design was safe but, wisely, recommended the addition of some superstructures to improve stability. The opinion of the public and of many sailors was still hostile to the design, not realising that without sails and with only light masts the heel induced, even by the most severe gale, was small and well within her safe range of stability.

As further re-assurance Froude carried out a series of tests in March and April 1871 on a 9 foot (1/36 scale) model of DEVASTATION, weighing just over 600 pounds. The model was first rolled in still water and then allowed to roll freely in waves in the sea off Portsmouth. It was tested with various sizes of bilge keels from nothing to a depth corresponding to six feet on the full scale.

In the still water tests, the model was pulled over to a given angle and then released; the number of rolls until the ship came to rest was then counted. The angles used were that at which the deck edge entered the water (8.5 deg) and that which just submerged the top of the armoured breastwork (24.5 deg). The results gave overwhelming evidence of the value of bilge keels; the number of rolls was reduced from 30 without keels to about 4 with keels six feet deep. (9 degrees with the 21 inch keels finally fitted) (ref 5.6).

The model was then taken into the sea where waves were found of similar period to that of the model. Trains of 4 or 5 waves were found, 15 to 18 inches high, crest to trough, corresponding to 45–54 feet for the full size ship with exceptional ones of over 70 feet equivalent. The angles were estimated by three observers, whose results agreed well, using a batten with one arm aligned on the horizon and another parallel with the deck. Further, similar tests were carried out with an 18 foot model of the same ship. In this second series, roll angles were measured continuously using an automatic apparatus. Without keels the average roll was about 20 degrees with occasional capsizes which was reduced to an average of 1–2 degrees by the six foot keels.

The remaining drawings of DEVASTATION show 21 inch bilge keels, probably fitted at a refit in 1880 (ref 5.7) By then Froude was dead and no trials seem to have been carried out. Model tests in waves were not carried out with the 21 inch keels but it is likely that they would have reduced the roll in the conditions described above to about 12 degrees.

5.5 Interpretation

Froude then devised a beautiful method of graphical analysis which started with a graph of the maximum angle reached in each swing against elapsed time; a 'curve of extinction' as he called it. Some fairly simple mathematics leads to a differential equation in which the rate at which the roll diminishes is expressed as an integral of the resisting force and the ship geometry which can be solved graphically.

Froude used this approach to compare and understand the results of the tests on the two models of DEVASTATION reaching some conclusions which surprised him. He found that the curve of extinction was virtually the same for the large model as for the smaller one and also that the tests made without bilge keels suggested that resistance to roll varied with velocity rather than with its square as would be expected if it were primarily due to friction. If resistance varied with the square of the velocity, the rate of extinction should be considerably greater in the smaller model and hence it was clear to Froude that some form of resistance other than friction was operating. He wrote in August 1872 (ref 5.8)

> 'On travelling again and again over the whole question the idea suddenly suggested itself that the waves created by the oscillation had been left out of account........ That some such action must take place is indeed obvious; that it does in fact take place, who can forget who as a boy has enjoyed the fun of wave making, by standing with his legs apart in a boat of manageable crankiness, and rocking her from side to side, to the imminent danger of himself and his companions, unless they have been swimmers?'

These waves take energy away from the rolling boat, the resulting changes in pressure round the boat forming a powerful resistance to motion.

This wave making resistance was quantified using some of the results of a trial with two sloops, GREYHOUND and PERSEUS, of 1100 tons, held off Plymouth in October 1872 with Henry Brunel in charge of the preparations (paid at 12 shillings per day). As with all trials, there was a great deal of preparatory work, Henry's letters mentioning the need to agree flag signals, the choice of pens for the recorders and teaching Froude's new Admiralty assistant, Phillip Watts, to use these pens. There was also a complaint that the dockyard had charged for 200 tons of ballast at £3-10-0 per ton (ref 5.9). Henry was susceptible to sea sickness and wrote that he didn't over-rate seasickness 'can manage pointer and vomit but couldn't do the pen business, vomit makes one lazy minded and overlook defects.' As an aside, modern research shows that rolling is unlikely to cause sea sickness which is due primarily to vertical acceleration.

Both vessels had a natural roll period of 4 seconds for a single roll, out to out. GREYHOUND, only, had temporary bilge keels 3.5 feet deep and 100 feet long fitted for the trial. The two ships were towed out to sea and placed beam on to waves of 4–5

Chapter 5: Rolling 1870–1875 : Part II – Application and Trials

second period. The GREYHOUND had an average roll of 6 deg, with a maximum of 7 deg while the PERSEUS, without bilge keels averaged 11 deg with a maximum of 16. The evidence was strengthened when GREYHOUND's rolling suddenly got worse as part of the temporary keel had fallen off.

Froude analysed the data from this trial (ref 5.10) and showed that the total work done on the ship by all forms of resistance was about 4,700 ft lbs. Some simple estimates showed that friction accounted for about 120 ft lbs and that from the keel and deadwood a further 700 ft lbs. He went on to calculate the size of the wave which would account for the difference. The full period of the wave would be the same as that of the ship (8 seconds), corresponding to a length from crest to crest of 320ft. The height then needed to account for the work unaccounted for would be a mere 1.25 inches, virtually invisible.

This incident shows Froude at his best: his tests on the DEVASTATION models were carefully observed but subsequent numerical analysis showed a considerable discrepancy. Deep thought suggested a possible explanation which, in turn, was tested by further calculation. Henry notes that in a later part of the trial, in December 1872, Froude hurt his shin climbing aboard the GREYHOUND and had to take a day in bed to recover. There were also successful trials of bilge keels in two Indian troopships, CROCODILE & SERAPIS.

5.6   The Measurement of Roll

The measurement of roll is not easy; Froude's 1857 experiments showed that the apparent vertical on a ship in waves is perpendicular to the wave surface. A simple pendulum at the centre of gravity will line up with the apparent vertical so that a very stiff ship, rolling with the wave surface will measure no roll on such a pendulum. Conversely, a ship which has very small stability and is hardly rolling with respect to the true vertical will seem to be rolling considerably if measured by a pendulum. Fig 5.2, below (ref 5.11), shows the difference between the absolute angle of roll from the vertical and the apparent or

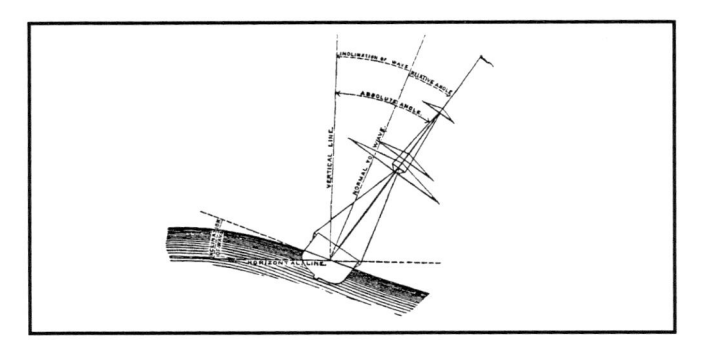

Figure 5.2 – Shows the difference between the absolute angle of roll from the vertical and the angle relative to the wave.

Figure 5.3 – Froude's roll measuring equipment. The large wheel forms the long period pendulum with the short period pendulum above the recording drum.

relative roll measured by the pendulum.

If the pendulum is not at the centre of gravity, its point of support is moving and there will be further errors depending on the distance from the centre of gravity to the pivot and on the period of the ship. As a rough guide, one might expect an error of about 20% from a pendulum on the upper deck of an ironclad and of some 50% on a wooden battleship.

The easiest way to measure true roll is to sight on the horizon, using the ratlines, a method used in Froude's earlier work. Obviously, this can only be used in daylight and when there is a clear horizon, unobstructed by wave crests. In early 1872 Froude designed an automatic roll recorder to measure both true roll and apparent roll so that wave slope could be deduced. This apparatus will be described as it is one of the earliest example of Froude's great skill in the design and building of accurate yet robust equipment. It remained in occasional use until the mid 1920s before being sent for honourable retirement in the Science Museum where it is still on show.

The recorder (fig 5.3) uses two pendulums, one of very short period (fig 5.4) which will hang perpendicular to the wave slope and the other of very long period (fig 5.5) which, for all practical purposes, will remain vertical as the ship rolls. The short period pendulum consists of a tube of brass, filled with lead, 2.5 inches in diameter and 20 inches long. It is mounted horizontally on knife edges with a period of .2 seconds. A very light arm from the pendulum carries a pen tracing the motion on a drum driven by clockwork. Another "clock" marks the paper at regular intervals as a time base.

The long period pendulum is a heavy wheel, three feet in diameter and weighing 200 pounds. This is carried on a steel axle, one inch diameter, arranged so that the centre of gravity is six-thousands of an inch from the axis giving a half period of 34 seconds, several times that of typical ship periods (ca 5–6 seconds) or that of most waves. Friction rollers support the wheel. It carries a wooden semi circle which, through a light rod, carries the relative motion of ship and wheel to the same drum. Froude was very proud of this machine and described it several times (ref 5.12).

In his first description he says, 'a weight of 1/250,000 part of that of the wheel itself, if placed at its extreme radius, would produce an oscillation of 1¼ inches in range, and which would continue for many minutes; or if the wheel was moved through 90 degrees from its position of rest, the oscillations would continue for nearly twenty minutes, the movement being so slow and solemn as to impress on the mind of an observer who had not seen it put in motion that the action was self-originated or induced by some mysterious agency.'

The curves drawn on the drum provide a continuous record from which can be obtained:

Figure 5.4 – The short period pendulum.

Chapter 5: Rolling 1870–1875 : Part II – Application and Trials

(1) The relative inclination of the ship and effective wave slope at any instant.
(2) The corresponding inclination to the vertical.
(3) The period of roll at any instant – seconds from out to out.

From (1) and (2) can be deduced the slope of the effective wave surface at any instant and the period of this effective wave. (For large waves, this effective period is virtually the same as that of the surface wave)

The roll recorder was modified by R. E. Froude for trials in the INFLEXIBLE in 1882 with knife edges in place of the rocker supports and a damping trough to reduce unwanted oscillations. It was used again in the trials of REVENGE in 1895 and during the tests of a gyro-stabiliser in VIVIEN in 1925 (ref 5.13). The short period pendulum was used more frequently and was given a clockwork timing mechanism.

Figure 5.5 – The long period pendulum now on display at the Science Museum.

5.7 Trials

A number of trials were carried out by Froude on the rolling of the DEVASTATION, partly to reassure the Admiralty, Parliament and the public of her safety and partly to confirm the results of the model tests. Henry wrote in April 1873 that it was gratifying to find how much he was appreciated, not least by naval people. Then, as now, ship trials were hard work, uncomfortable and great fun and it is hoped that the following account, based on Henry's letters as well as the reported results illustrates all three aspects.

In the first series, April–May 1873, DEVASTATION was rolled in still water by the action of men running backwards and forwards across her deck. This is not as difficult as it sounds and, provided that a clear deck space is available, a large angle can be built up quite quickly and measurements taken as the roll decays with the men at rest. This technique was still in use well after World War II. The trick is that the men must run from the middle line to the side and back in half the natural period of roll of the ship. In practice, it is self adjusting with the men told that they must always run 'uphill'.

In the case of DEVASTATION, a roll of 7 degrees was built up by 400 men running 18 times across her deck. The original plan for DEVASTATION was to use 560 men, four deep, for which it was estimated that a length of 280 feet would be needed (ref 5.14). A glance at the illustration of the ship (fig 5.1) shows that this must have been very difficult since the ship was only 285 feet long and the deck much obstructed. Henry describes the first trial in a letter of 11th May 1873 –

'Rolled in harbour with 200 men. Froude let a commander give the word to run and so was made a mess of, then HB did in proper style but not enough men; will do again with men from other ships but stubborn to roll...

DEVASTATION went to sea in April, mainly as a first check that all was well but Froude and Henry were on board and there are some interesting comments from Henry's letters. Froude had the captain's spare cabin which was behind armour and very gloomy. Henry and Phillip Watts, an assistant constructor lent by the Admiralty to help Froude, shared a 'daylight' cabin in the superstructure. It was under the after gun turret and Henry hoped that they would not fire at night as 'the shot will pass over HB and frighten away what wits he has.' (ref 5.15). He was busy collecting bits of apparatus which had been forgotten and notes that there was no 'po' in his cabin so went ashore and bought one. Froude ate with the captain (Hewett) while Henry and Phillip messed in the wardroom. Henry says the captain was very pleasant but only the navigator and one or two of the sub lieutenants were really interested in science. However, in another letter, Henry describes the officers as 'kind, agreeable, cordial and wondering'.

Even today, trials parties are often seen as a nuisance and their reception on DEVASTATION seems quite favourable. .Froude, always got on well with people and both he and Henry were careful not to rush their work or be too demanding on help from the ship and also taking care to site their apparatus where it would not be in the way.

The sea was calm during the six hour trial and Henry notes that he was not sick with the ship only rolling three degrees in the barely perceivable swell. Running with the sea on the quarter gave a period of encounter with the waves of 15 seconds, near enough to the ship's own period of roll to give some movement.

DEVASTATION went to sea again in August and September 1873 looking for bad weather and based on Berehaven (Cobh). She was in company with the older ironclads SULTAN and AGINCOURT, the former having a broad white stripe painted on her side at the height of DEVASTATION's upper deck to help in comparing the motions of the two ships. On September 19th she went out with SULTAN into a 45 knot gale with 16 foot waves, 400 feet long. Head to sea at 7 knots, the DEVASTATION pitched 7 degrees, out to out, a bit less than SULTAN. However, her low fore deck was swept by green seas, as was intended, since these seas acted to reduce her pitching. Placed beam on to the sea, neither ship rolled severely, the period of the waves being only about half that of the ship.

There is a nice story of this trial, illustrating Froude's popularity with the navy. When DEVASTATION rejoined the rest of the squadron, the band of the SULTAN, on which Froude had previously carried out trials, struck up a music hall ditty of the day – 'Willy, we've missed you'.

A week later DEVASTATION went out with AGINCOURT in

Chapter 5: Rolling 1870–1875 : Part II – Application and Trials

waves 450–600 feet long and about 20–26 feet high. The older, longer ship pitched less, DEVASTATION averaging 5–8 degrees (11.75 max). The greatest roll angle was at 7.5 knots with the sea on the quarter with 13 deg to windward and 14.5 to leeward with 5–6 degree of pitch which Henry described as follows:

> 'Grand pitching and plunging yesterday, taking in green seas.'

The low freeboard forward was not a success, particularly as the deck leaked and had to be caulked with tallow. A description reads:

> 'A wall of water would appear to rise in front of the vessel, and dashing on board in the most threatening style as though it would carry all before it, rushed aft against the fore turret with great violence, and after throwing a cloud of heavy spray off the turret into the air, dividing in two, to pass overboard on either side.'

Froude and Henry rejoined in November 1874, Henry says that his cabin was comfortably arranged but the bed was difficult to fold up and it was very cold. The ship's plans were hung up outside and he could hear sailors explaining the ship to friends but 'doesn't hear anything against the safety of the ship.' The 'bogs' stank again. After five weeks preparation, the Admiralty decided to postpone the trial to the, spring. Henry thought the reasons were contemptible but

> 'the truth is they are in an ignorant funk about her and the same ignorance that let the CAPTAIN be built is acting against the DEVASTATION. 'The Constructors' Department are mad about it but 'old' Admiral Milne, now First Sea Lord was the Admiral of the fleet when CAPTAIN was lost.'

When the trials team left, Henry told Froude that there was no need to give champagne to the wardroom as they all liked him so much and there was no need to use gifts to gain gratitude.

The trial was finally held in April–May 1875 which was more enjoyable. Henry joined the ship at. Spithead, together with Admiral Stewart, Controller, Nathaniel Barnaby (the Chief Constructor) and Captain Boys. This time Henry wrote to his mother that the preparations went well; his cabin was comfortable 'bigger than the Nile boat' and he could write comfortably as everything was within reach. They arrived at Plymouth on the 14th for what seems to have been a hilarious meeting. DEVASTATION was still seen as potentially dangerous and in need of escort. The original choice was HERCULES but she was delayed by a small fire and UNDAUNTED, a wooden screw frigate, was chosen to take her place. She was the flagship of Admiral Mc Donald who was "Cussing and blinding" over the possible delay in his voyage to the West Indies. He had no faith in steam and was sure that at least one ship would break down. Eventually it was decided to wait three days for HERCULES and Controller and his team visited the tank at Torquay to see Froude's experiments.

McDonald's contempt for steam was not unique but was much less common than is usually claimed. Henry, giving his usual snap judgements, liked Admiral Stewart and found Boys 'a very willing listener to words of wisdom.' Barnaby was 'too much in the habit of giving any answer that will sound like an argument. Dealing with Admiralty Boards probably begets this.'

Froude joined the ship when they sailed for Lisbon on April 17th. Henry wrote,

> 'Abominably fine weather, so no chance for work except a few diagrams....... Apparatus working well and dodges for electric connections answer well in practice.'

By the time they reached Lisbon on the 22nd they had no useful results though Henry noted the excitement when 'all four guns were fired at once by electricity which had a stunning effect.' (They were 12 inch rifled muzzle loaders firing a 706lb projectile with a charge of 110lbs of powder – stunning indeed.) They also tried to measure wave length using a float on the end of a long line. With the stern on the crest of a wave and the float on the next crest, the length of line paid out corresponded to the wave length but it was difficult to see the float and the scheme was not successful.

The King of Portugal visited the ship on April 24th and 'Mr Froude's apparatus was almost the principal show thing, and as a show off we rolled the ship by running men from side to side for the King to see the apparatus work, and then he came on deck to see the men running.' Later the King inspected Froude's rolling diagram: both he and the Portuguese Admiral had read Mr Froude's book so 'Froude was certainly a prophet out of his country whatever he may be in.' In another letter, Henry says the King 'couldn't be got out of the ship, he liked it so much.'

On the way to Gibraltar, they met

> 'a low co-periodic swell, only 2 degree slope of wave, but we got once nineteen degrees of arc... We got nice and useful diagrams, but of course not at all as much as we want.'

These were analysed on arrival at Gibraltar and will be discussed after the description of the Froude-Brunel 'holiday' is complete. They reached Malta in DEVASTATION on May 5th and were entertained by the governor and the admiral 'Drummond had gout which supplied conversation'. The trip had been 'calm as Torbay when fine weather'. After sightseeing and buying some souvenirs, they voyaged to Brindisi in the admiral's yacht HELICON – 'flat blue sea with a bright blue sky.'

During the following years, Froude was more involved in propulsion work than in rolling but there were a few more trials, notably with SHAH in 1875. In January 1875, Henry wrote that SHAH was to be fitted with bilge keels 'which is a triumph for Froude's principle.' In July he wrote 'First rate trial

Chapter 5: Rolling 1870–1875 : Part II – Application and Trials

of SHAH after bilge keels put on.' There is an interesting sidelight on Froude's tact in another letter. Someone had erected a 'pulpit' for the trial – 'just the thing Froude disapproves of, to do conspicuous things needlessly, wishes held have taken down and put away till needed.' Also in 1875, Froude, reporting rolling trials on the FANTOME, refers to a diagram, devised by Henry Brunel, showing the period of encounter with waves for all course angles.

Then in July 1877, the Admiralty set up a Committee to examine the stability of the new battleship INFLEXIBLE (ref 5.16). The chairman was Admiral Sir James Hope and the members, all engineers, were Wooley, Rendel and Froude. The secretary was always one of Froude's assistants, first J. R. Perrett, then P. Watts and finally Edmund Froude.

The ever increasing size of guns – 16 inch muzzle loading rifles in INFLEXIBLE – necessitated ever thicker armour which, because it was so heavy, could only cover a part of the length. INFLEXIBLE's 24 inches of armour weighed, with its teak backing, 1100 lbs per sq ft. Reed, who had left the Admiralty over the CAPTAIN affair, attacked the design, suggesting that if the unarmoured ends were flooded in battle, the remaining stability would be inadequate. The investigation was extremely thorough and included a number of experimental and theoretical investigations by Froude.

Several of these investigations used an elaborate model, built at Portsmouth, 1/24 full size and weighing nearly a ton. The armour deck and citadel bulkheads were watertight but the 'unarmoured' ends were made of a grid of iron bars while the bulkheads in the unarmoured area were of wire mesh. The cork fitted for buoyancy in the ship was reproduced in the model and could easily be removed.

The model was rolled in various conditions while at rest and, to most people's surprise, roll damping was much greater with the ship partially flooded. For example, when the ship was released from rest at 10 degrees heel the next roll would be 9 degrees with the ship intact and between 2.2 and 5.7 degrees with various flooding at the ends. In fact, the water in the flooded ends acted to oppose the rolling.

There was also concern that INFLEXIBLE might plunge by the bow if moving at speed with the ends flooded. The model was brought to Torquay in the tug MALTA and a letter of 9th October 1877 refers to its transfer to the 'Admiralty Experiment Works', the first time this title was used for Froude's tank, a title it was to retain for more than a century. The model was towed down Froude's model tank at Torquay in various flooded conditions, During this test special pieces were put in the side representing plates hit by shot and forming scoops. Finally, Froude tested models representing a ship similar to INFLEXIBLE but with increased beam and finer ends. It was found that such a form could reach the designed top speed with less power whilst the increased beam gave more stability. (See also Chapter 13.)

Understandably, the committee reported that INFLEXIBLE was safe, even under the most severe conditions of damage. With hindsight, they were probably wrong. She had a large number of watertight doors leading into the central citadel and experience shows that such doors almost invariably leak under severe damage which distorts the structure. The model seems to have been used for instructional purposes at the RN College until it was scrapped during World War II.

Even in 1872, Froude told the Royal Commission on Scientific Education

> 'But I must add, the advance thus made appears to me to be not quite so complete in relation to the principles which govern the rolling of ships; they do not seem to me to have been quite so scientifically applied even in the ships which have most recently been constructed.'

This lack of understanding of his rolling work and hence lack of application was to continue. In 1889, that great naval architect, Sir William White, designed the ROYAL SOVEREIGN without bilge keels thinking that the great inertia of this ship would limit rolling. Soon complaints were made of very heavy rolling and REPULSE was fitted with keels 200 feet long and three feet deep. In comparison with a sister ship, she rolled 11 degrees whilst the other vessel rolled 23 degrees. Even in World War II, the CAPTAIN class frigates had to be taken out of service for bigger bilge keels and at least two classes have needed modification in more recent years. Froude showed how this unnecessary discomfort, even danger, and the resulting cost could have been avoided. Very recently Monk showed in a graduate study how to get bilge keels right at the design stage (ref 5.17).

Froude wrote to Barnaby on 25th March 1873 (ref 5.18) expressing his disquiet over another aspect of rolling work. For many years it had been customary, when a number of ships were together in rough weather, to record the roll angles of each ship over a period of five minutes, reporting the average and maximum roll in this period. In his long letter which can only be summarised, Froude said that such trials were of little scientific value as different ships had quite widely varying natural periods of roll and would respond to different wave lengths. In irregular waves, it would not be possible to identify predominant frequencies in a time as short as five minutes. If the wave crests were not parallel, the difficulties would be much increased. He would prefer to measure ship response to carefully selected groups of waves whose features and effects would be noted by several officers who would estimate the height, length and period of each individual wave and note the

Chapter 5: Rolling 1870–1875 : Part II – Application and Trials

roll angle each time. 'Competitive' trials as then conducted were 'a game of chance'.

References

5.1 K. C. Barnaby. Some Ship Disasters and their causes. Hutchinson, London, 1968.

5.2 D. K. Brown. The Design and Loss of HMS CAPTAIN. Warship Technology 1/1989, London, RINA. Also D. K. Brown. WARRIOR to DREADNOUGHT. Chatham, London, 1997.

5.3 D. K. Brown. British Battleship Design 1840–1904. Interdisciplinary Science Reviews, London, March 1981.

5.4 N. A. M. Rodger. The Admiralty. Terence Dalton, Lavenham, 1979.

5.5 As 5.2.

5.6 William White. Manual of Naval Architecture. John Murray. London, 1900.

5.7 Letter to the author from the National Maritime Museum, 1993. (G. Sattler)

5.8 W. Froude. On the Influence of Resistance upon the Rolling of Ships. Naval Science, October 1872. (p.155 in collected works.)

5.9 Brunel Collection, August 1872.

5.10 As 5.8.

5.11 W. Froude. On the Rolling of Ships. Trans. INA., London, 1861. (Collected works)

5.12 W. Froude. Three papers, all contained in the collected works.

Description of an Apparatus for Automatically Recording the Rolling of Ships. British Association, 1872.

Same title, Trans. INA., 1873.

Apparatus for Automatically Recording the Rolling of a Ship in a Seaway and the Contemporaneous Wave Slopes. RUSI 1873.

5.13 R. W. L. Gawn. Historical Notes on Investigations at the Admiralty Experiment Works, Torquay. Trans. INA., London, 1941.

5.14 As 5.9, 19th April 1873.

5.15 As 5.9, 24th April 1873.

5.16 Report of the Committee on the Inflexible. London, 1878.

5.17 K. Monk. A Warship Roll Criterion. Trans. RINA., London, 1987.

5.18 Froudes' Museum.

The Way of a Ship in the Midst of the Sea

Chapter 6
A Sacred Duty to Doubt

Froude and Newman, The Nature of Proof in Religion

6.1 Introduction

The early relationship between William Froude and Newman has been covered up to Catherine's conversion in 1857. This chapter will deal with the philosophical debate between the two men on subjects such as the nature of proof, doubt and certainty, and scientific method. The crisis caused by Edmund's expressed wish to become a Catholic priest is discussed as are the happier problems concerning Newman's dedication of a book to William, both showing the strength of feeling on both sides. Though this chapter appears to separate Froude's religion from his scientific work, the views which he put to Newman throw much light on his way of working whilst the upsets of family life must have affected his professional work.

In a long letter to Newman of 29th December 1859 (ref 6.1) Froude explains how his thinking on religious matters developed.

> 'By slow – (but only by slow) degrees the convictions I refer to became masters of my whole mind, mastering the dogmatic habit of thought, first in relation to professional knowledge and scientific enquiry (for when I was an undergraduate the general tone of Oxford teaching was at least as dogmatic in relation to sciences as in relation to Theology and had laid strong hold on me, there), and then penetrating at length into the region of Theology and altering my views in relation to it, so as to produce results which I fully admit to be at variance with Hurrell's direct teaching.'

By the early 1850s, Froude's views seem well established and with one or two exceptions did not change very much. In consequence, appropriate quotations from quite different dates have been juxtaposed to illuminate the continuing debate with Newman. The selection is, of course, that of the author and may, despite care, still reflect his own prejudices.

Froude's religious observance continued to follow Anglican tradition. Harper (ref 6.2), presumably quoting Eliza, says that he continued to discuss religion with his family and discuss its relation to science. He appreciated the beauty of the Anglican service, both the music and the language, and knew by heart many of the poems of the 'Christian Year'. He read the Bible to his family, seeking out the meaning of the most beautiful passages.

Figure 6.1 – John Henry Newman as Cardinal, after Froude's death.

6.2 Faith and Proof

Faith

Newman's most succinct definition of Faith is given in a letter to Catherine Froude (27 June 1848)

> 'It is, we know, the Gift of God, but I am speaking of it as a human process and attained by human means, Faith is thus not a conclusion from premises, but the result of an act of the will, following upon a conviction that to believe is a duty'.

Froude's clearest answer to such a viewpoint comes in his letter of December 1859, quoted above. He says:

> 'I do not overlook the view that 'Spiritual insight is granted as the reward of Faith," nor do I venture to judge that (in some shape) it is an impossible or even an improbable one. Yet I feel it to be one in the highest degree improbable if the merit of Faith is measured as the Theologians seem to measure it, directly as the positiveness of the Belief and inversely as the strength of the evidence. Thus measured Faith seems to be but another word for 'prejudice' – i.e., as the formation of a judgement, irrespective of, or out of proportion to the strength of the evidence on which it rests and I regard it as an instance of an immoral, temper or the immoral use of the faculties.' – to Froude 'immoral' was a very strong word.

Froude's attitude to Faith also had a positive side to it and he continues:

> '...the only pattern of Faith which I can conceive to be meritorious, is the temper which, while it realises as carefully as possible the exact degree of doubtfulness which attaches to its conclusions, +acts nevertheless confidently on the best and wisest conclusion it can form –..'

Froude then suggests that this approach to faith has only recently been applied to the 'pursuit of scientific truth and in the cultivation of the mechanical arts' which in consequence, have 'made progress with increasing rapidity and security' while the principle has even made progress in politics.

Newman saw proof in religious matters being almost instinctive and depending on what he called the 'illative sense'. 'All men have a reason' said Newman, 'but not all men can give a reason.' Harper says that a favourite illustration of his was of the weather-wise farmer predicting rain when every perceptible sign pointed to a clear day. He pointed out that geniuses have often been noted for this ability to overgo the merely logical faculty of the mind and arrive in a flash at the truth. It was by such flashes that Napoleon could with a glance take in the disposition of enemy forces and have immediately a counter-plan in mind. In mathematics Newton had the same kind of intuition. Newman's whole argument depended on the validity of this instinctive ability to arrive at right conclusions through the 'accumulative force of a multitude of reasons which if taken singly were only probabilities.' To this instinctive faculty he gave the name of the illative sense.

Newman was also found of using geographical analogies to represent things which were certain. For example, in his last letter (29th April 1879) he wrote 'Then I go on to say what scientific men believe of Great Britain, viz its insularity is an absolute truth, (cf probability) that we believe of the divinity of Christianity'. In the same letter there is an interesting justification for the foundations of Catholic doctrine.

> 'As Newton's theory is the development of the laws of motion and of the first principles of geometry, so the corpus of Catholic doctrine is the outcome of Apostolic preaching. That corpus is the slow working out of conclusions by means of meditation, prayer, analytical thought, argument, controversy, through a thousand minds, through eighteen centuries and the whole of Europe.'

Scientific Method

Newman referred several times to Newton's theory of gravity as something which could be believed with certainty, for example, the following extract from a letter of 10 April 1854.

> 'Nor do I think it matters much that many men are 'certain' of what is opposite to Catholic faith – or 'certain' that Catholicism is false – for men have been 'certain' that Newton was false – yet that would not move me against Newton, because, though we are no judge of Newton's reasonings, we may be judges of persons who judge and embrace them, ...'. Froude dissented, a later reference (25th January 1860) making his point most clearly.

> '...it seems to me that you attribute to scientific proof a cogency and completeness of conviction, which in the domain of 'science' technically so called, which none of the higher minds which occupy that domain, attribute to such proofs. I have seen either in your letters or in something which you have published more than one illustrative reference to the Newtonian system, as if it were to be treated as being established beyond the possibility of confutation or of change. And in your last note, you refer to the proof of the earth going round the sun as an illustration of the sort of 'proof' by which Christianity is proved. It is of course difficult to understand exactly how strong or complete a meaning, another person attaches to the word 'certainty' – but there is a test which seems to me effective as showing the limitation of its meaning which the best men of science adopt. They would all emphatically disclaim a such a certainty as would justify them in saying, 'I will always hold to this – I will earnestly endeavour to combat arguments brought against it and will resist the temptation to be swayed by them.'

Froude is making two different points, one that Newton's theory is not 'certain' and may need to be revised – as it was when Einstein's theory of relativity was accepted – and secondly that acceptance of the theory was not an effort of will but depended on its consistency with all available evidence.

Froude was also unhappy with the idea of scientific progress coming in the form of a sudden 'illative' flash (29th June 1853)

'Science progresses only by following clues – it is only when, in such a pursuit, our stock of acknowledged principles fails to account for some unmistakeably residuary phenomenon, that experimentalists venture cautiously to think that they have really got hold of what may turn out to be a new and unacknowledged principle. In a little book ....it was said, I remember, that all the great principles of science had come into men's' minds heaven born as it were and I remember that my mind misgave me, even while I assented and admired the proposition...... now I don't assent to it; generally speaking the great scientific discoveries are but the crowning results of some series of long continued and patient enquiries of men known as other enquirers ....and whose minds are for the most part as well disciplined to perceive the bearing and the upshot of their experimental facts as to decipher the experiments and see the facts which they present –'.

This view of scientific discovery will be compared with Froude's own account of his 'sudden' discovery of the importance of wave making in ship roll damping in the final chapter.

Doubts

Froude's views are most clearly expressed in his letter of 29th December 1859.

'Our "doubts" in fact appear to me as sacred,...... more strongly than I believe anything else I believe this – that on no subject whatever, distinctly not in the region of the ordinary facts with which our daily experience is consonant, distinctly not in the domain of history or of politics, and yet again a fortiori not in that of Theology, is my mind (or as far as I can take the mind of any human being), capable of arriving at an absolutely certain conclusion.'

Newman's views are drawn from his letter to Catherine of 10 April 1854.

'I do not see then that I am bound to believe WF's statement of the unsatisfactoriness of religious inquiry, and the necessity of an everlasting suspense, until I am sure that he contemplates the probability of that being true, which is not improbable in itself, and which all those who have attained certainty say is true – that a preparation of mind of a particular kind is indispensable for successful inquiry – and till he makes it clear to me that he duly appreciates that probability.'

Froude's response was:

'I know that all the really high cast minds, which are engaged in the advancement of science and also pursue it in that really philosophical spirit which alone serves to consolidate the advances made, all treat their own conclusions with a scepticism as profound, and as corroding as that with which they treat Theology. The scientific principles which are regarded as most certain, are those the probability of which +is being most continually tested and found to stand the test.'

Froude's hydrodynamic work provides many examples of his approach, doubting even his own results and conclusions, in a constructive way, so reducing the areas of uncertainty.

Uncertainty

'To us, probability is the guide of life.'

The phrase above is often used by William Froude in his letters to Newman, with small variations in the wording. It originates in Bishop Butler's 'Analogy' but was picked up and frequently used by Hurrell Froude. William would not accept any rule as 'certain' – see his comment on Newton, earlier. Newman felt that some of there differences were merely semantic:

> 'We differ in our sense and use of the word "certain". I use it of minds, you of propositions. I fully grant the uncertainty of all conclusions in your sense of the word, but I maintain that minds may in my sense be certain of conclusions which are uncertain in yours.' (29th April 1879)

Sacred and Secular

Newman saw a real difference between sacred and secular matters with a corresponding difference in the way such propositions were proved but Froude pointed out the difficulties in this division.

> 'At least, it seems to be a tenable view, a priori, that men are intended to deal differently with their conclusions when these lead up into Religion from that way in which they deal with conclusions relating to the ordinary affairs of life – as if instinct were to guide them in one case, logic in the other; though an abundant crop of intractable difficulties arise when one attempts to reduce the distinction into a rule of practical application, and though your grounds of demur seem to make themselves felt towards the view itself, when its corollaries are looked into. For in the first place, it is extremely difficult if not impossible to draw a clear and available line between the probabilities which lead up into Religion and those which belong to Common life so interwoven the two classes of questions are, when one really looks into them.' (8th October 1864)

Last Words

During his last voyage to South Africa, Froude wrote an extremely long letter to Newman giving his views on religion. Newman received this letter on 1st March, just as he was preparing to travel to Rome to be made Cardinal and in acknowledging the letter asked if he might keep it a little longer so that he could make a full reply. Froude's letter was returned to him on 16th April and only remnants of the draft have survived. Despite the fact that it was such an exhausting time for him, Newman drafted very extensive notes and dated a full reply on 29th April but news of Froude's death arrived before it was sent. An extract from Froude's letter follow; extracts from Newman's reply have already been used.

> 'I have said that so far as I can see it is only out of the consciousness and sense of duty owed that we can rightfully construct the idea of a divine Person; resting the opinion on the perception that the notion of a debt no less naturally involves the notion of a Person to whom it is owed than it involves the notion of him who owes it. The idea of a divine Person does not, to me, rightly to grow out of the perception of the existence of laws by which the course of nature is governed. I cannot feel that the law implies a personal Law giver in the same way that dept implies a Person to whom the debt is owed.'

This passage suggests that Froude was moving further still from Newman's views and perhaps moving closer to atheism.

6.3   Significance

This long and detailed correspondence shows much of the way in which Froude's mind worked, his views on scientific method and on the nature of proof. As an aside, one may quote from his letter to Henry Brunel of 15th May 1857 in which he says; '...the time spent in correspondence is double time, as it involves both the letter and replies.' Religion was clearly an absorbing topic to Froude shown both by the time which he devoted to this correspondence and by his very strong reaction to the conversion of his family, discussed in the next section. The long debate with Newman led to a hardening in Froude's attitude and increasing doubt over many aspects of belief. His views on Roman Catholic doctrine were not put very forcefully in his letters to Newman, possibly to avoid offending his old friend, but in the discussion over Eddy's vocation for the priesthood, which follows, it is clear that he saw that doctrine as incompatible with scientific method.

The value of the debate to Newman is less clear. Froude was just one amongst a very large number of regular correspondents, with which he exchanged innumerable letters on all aspects of belief and, in consequence, many of Newman's biographers do not even mention William Froude. Certainly, Newman took their correspondence seriously as Froude's letters are often marked with marginal notes for reply. However, William was almost the only friendly critic of Newman's thinking and representative of the new breed of scientific doubters and as such must have been of great value to Newman, a view strongly supported by a letter of 18th January 1860. 'I should ask your leave to put various points before you as iron girders are sent to the trying house.' Newman had a number of devoted supporters and many hostile critics; Froude was his only critical friend.

To a great philosopher such as Newman, the value of such severe but helpful criticism is invaluable and justifies the view that William Froude, like his elder brother, Hurrell, was indeed an important influence on Newman. It would seem that important parts of the 'Apologia' are drawn from Newman's correspondence with the Froudes and particularly with Catherine whose importance will be brought out in the next section. Though the 'Grammar of Assent' had its origin as a response to F. Stephen's 'Dr. Newman's Apologia', it was both suggested by Froude and was, to a considerable extent, based on their correspondence.

6.4   Conversions

As already described, Catherine was converted in 1857. The Froude's eldest son, Hurrell, had also been corresponding with Newman and as a result, was received by Newman into the Catholic Church on 24th December 1859. Newman wrote to William the same day to inform him but also challenged Froude's own beliefs.
'As to yourself, I do not believe, I never will believe, that in the bottom of your mind you really hold what you think you hold, or that you master your own thoughts. I think some day you will allow the truth of what I say. Accordingly, whatever pain it is to me to think of our actual differences of opinion, I feel no separation from you in my heart, I, please God, never shall.'

Newman took great care over this short letter, realising that the news would cause distress, and the draft is much amended. It drew from William a very lengthy response, from which several passages have been quoted in the previous section, setting out his views.

William was far from convinced that Hurrell understood what he was doing and thought he might return to the Anglican Church later. Newman's response was friendly but robust 'I meant that what influenced Hurrell was faith, not a fancy. On the other hand, I think who could no more promise that he would examine in the future than he could decently, becomingly or dutifully promise that his mind should ever be open to evidence that his parents were impostors and hypocrites. ..... Forgive me if there was anything rude in my last paragraph.'

William's reply was dignified:
> 'So long as my children think of me fairly – and weigh evidence about me fairly – and do not go out of their depth in inquiry – so long as they in conducting that particular enquiry all the care

which the pursuit of truth (the honest pursuit of it) requires of the pursuer everywhere – I see no reason to complain of their being ready to receive evidence that I am an impostor and a hypocrite. If I am such, I desire that they should know it – or if I flinch from thus saying here, it is only as I flinch from it while my better sense tells me I ought not to flinch. But I feel I have the right to require of them that they should not select the evidence which is against me – that they must not be suspicious.'

Again there were practical difficulties, centred over family prayers. Hurrell was lodging with William Fishburn Donkin, Savilian Professor of Astronomy at Oxford. Hurrell seems to have continued to attend the Donkin's family prayers but not participating which caused offence. Newman and William eventually found a compromise and Hurrell graduated, going to India as a civil engineer.

It is likely that the Froudes' oldest child, Eliza Margaret was converted at about the same time as Hurrell as her inclination to Rome was evident in her letters to Newman as early as 1857. The third offspring, Arthur Holdsworth, a naval officer, followed in 1861. Neither of these apostasies caused any great distress.

6.5   Edmund

The next son, Robert Edmund (ref 6.3) was to become his father's assistant and later his successor and, as such, must be considered in more detail, particularly as the correspondence sheds much light on the personalities of William and Catherine Froude and on their friendship with Newman. Edmund went to Heavitree School, near Exeter, from 1856–58 and then to Bradfield School till 1863. In 1862 he was engaged in correspondence with his mother concerning his religious beliefs, letters which Catherine passed to Newman, but not, apparently, to her husband. By the end of the year, both Catherine and Newman persuaded Edmund to let his father know of his belief. William was greatly upset as he saw Roman Catholic faith as contradictory to the spirit of inquiry which governed his own life and which he had thought was developing in Edmund.

Edmund became a Catholic on 9th April 1863, announced in a long letter from Newman to William summarised below.

> 'I have received Eddy into the Catholic Church today. He made it clear to me, that for some months you have been aware of his intention of being received, and of being received at this time. If he has to be received, I felt that you would rather I received him, rather than another.'

Newman rejects any suggestion that Edmund was not mature enough to make such a decision saying 'I believe him to be acting deliberately, on right motives, and on rational grounds,'

Chapter 6: A Sacred Duty to Doubt

He continues that he is not 'putting himself under a sort of intellectual tyranny by doing an act which he is not allowed to reverse.' There are indications, which will appear later and, mainly from Newman's own letters, that Edmund was still immature.

Newman clearly understood William's distress writing to Sister Mary Gabriel:

> 'I am engaged just now in receiving one of the Froudes – a boy of 16 who arrived here yesterday from the School. My dear friend, his father, who is not a Catholic, has seen his children one after another, (this is the fourth) received into the church; and he has borne it so gently, so meekly, so tenderly, (though it has given him a sense of desolation more cruel to bear) that I do trust God's mercy has the same gift in store for himself. .... What a good Catholic he would make if the grace of God touched his heart.' (7th April 1863)

It is interesting that several of Newman's letters envisage the possibility of William turning Catholic whilst William saw it quite possible that his family, if not Newman himself, might leave that church and return to the Church of England. It would seem that both men underestimated the strength of other's conviction.

Henry Brunel, already a close friend of all the Froudes, expressed some interesting and rather different views in an entry in his private journal for 1st February 1863 (ref 6.4).

> 'A very nice long letter from Mrs Froude which see. It certainly seems a most trying position for her though I question whether a more thorough openness is not more desirable – Ought she to conceal anything from Mr. Froude. Did she lead Eddy to the desire to join the Catholic Church? Did she not at any rate know that he was being so led – What is to be done in a such a case – Believing as she does I suppose she is right in placing her children's' souls' welfare before her husband's happiness but why should a change of religious opinion cause a separation between father and son and such that Mr. Froude should think he has "lost his sons". This is not right – Is the fault Mr. F. – '

Religion did not interest Henry very much and he may not have realised how much it mattered to his friends, the Froudes, though there seems justice in his suggestions that Catherine was unduly secretive and that William over-reacted.

William's letter of 25th April thanked Newman for his understanding of those whose views were so different from his own. Edmund left Bradfield School and entered the Oratory School. The fees were 80 guineas a year or £160 in the "Woolwich class" which Newman hoped he would join. In July, Newman reported Edmund's progress to Froude. He had done well in maths and in Latin and had started Greek. His conduct was 'a pattern to the school' and the letter concluded 'Eddy is a boy but a good boy – or, a good boy, but a boy' (He was almost

17, one of several indications of immaturity in Newman's letters)

William wrote again to Newman on 14th January 1864 saying that Eddy had a growing understanding of maths and:

> 'He is very quick and is solidly intelligent and this shows itself with special clearness in relation to any engineering work, measuring, planning and the like. Of this we just now have to do a good deal, as we are going to build ourselves a new house, and have a number of men at work levelling and shaping the ground –'

William asked Newman if Eddy could stay on at home for a few days longer to help with the building of their new home at Chelston Cross and:

> 'The more so as one of the Brunel's is (I hope) to be here just at that time, and he also has been giving so much assistance about our plans he and Eddy will be of great help in the consultation about many points which remain to be decided –'

Later it was agreed that Eddy should leave school in the summer and help with building the house. A tutor would be engaged to keep up his studies until he went to University, (Both Merton and Balliol were mentioned) at the end of 1864.

The youngest child, Mary Catherine (b 1848) had developed the family disease of tuberculosis, the illness getting worse through early 1864, until she died in her father's arms, aged 16, on 30th May. On 31st May William wrote to Newman with news of her death with harrowing detail. He also said of his last Protestant child

> 'There are many points in the history of her mind, and her views, and her character on which I could have wished to compare thoughts with you – in respect of which though I should express myself differently from you. ... But I cannot write now on such topics.'

Catherine also wrote to Newman on the day of her child's funeral (4th June 1864) saying:

> '... now that they have carried my darling child to Denbury, I am left alone in the house, and I feel that nothing could relieve me so much as writing to you. All Monday he was very much overwhelmed; that last terrible half hour, during which he held her in his arms, was always present to his imagination.' Edmund who was present at the death bed was greatly distressed.

In July 1864, Edmund, who had left school, returned for a retreat under Father Suffield. As a result, Edmund was led to believe that he had a vocation for the religious life and decided he would become a priest. Newman does not seem to have been aware of what was going on and was not pleased. This led to a correspondence involving Newman, Eddy, Catherine and William. There was a great deal of duplication and all wrote at great length so that their letters can only be summarised. However, these letters throw much light on the characters of

Chapter 6: A Sacred Duty to Doubt

those concerned.

Newman had originally arranged with Edmund that the latter should tell his father and that Newman would follow with a letter but, on reflection, Newman realised that William should be prepared and wrote on 19th July to warn him. Edmund had told Newman that he had had the idea of becoming a priest for some time and would have discussed it with Newman had the topic not arisen with Suffield. Newman made it clear to William that he considered Edmund too young to have a clear understanding of what he was proposing to undertake and thought Edmund should go to Oxford, if such was his father's wish, first.

William wrote back on the 20th.

> 'You could not have written more kindly. But what you tell me is a great blow to me, one of the blows which make me wish at the moment that what is to be done be done quickly.'

'In one side of Eddy's mind there was growing up a fund of common feeling and interest with me, and this seemed to give one little spot of light, and this intention quenches it by what seems to me (though I can well understand how it will seem widely different to others) not an impulse really divine and from above. But it is useless giving vent to what on my part is hasty and impulsive feelings.'

Catherine was also upset and regretted Eddy's action and wrote to Newman:

> 'Indeed the dear man (William) has had blow upon blow since first I announced my intention of becoming a Catholic – and this last seems the heaviest of all – for he had just begun to find Eddy a thoroughly pleasant companion and with Eddy's excellent professional promise Wm felt he might look forward to his being a distinguished person. All this hope is now dispelled; and he has the additional pain of feeling that Eddy has left a career of honour and credit for one which (I fear) is looked on by him with dislike and disapprobation – and which he feels must place a gulf of separation between them for ever.'

> 'This is all the more painful because evidently Wm had been the more agreeably surprised by finding Eddy's becoming a Catholic has not in any degree produced the separation which he anticipated. On the contrary they have been far more drawn together than they were before:– and Wm has often expressed himself to me as surprised to se how quickly Eddy catches his ideas and masters them – and often suggests improvements which had not occurred to Wm but which he sees to be judicious.'

> 'Wm is going to Salcombe to see Anthony, who is there, and that will do him no good'.

Newman wrote a long letter to Edmund on 24th July advising patience, and making clear the difference between a religious life and a vocation for the priesthood. He concluded:

> 'I should recommend you going about your direct duties whatever they are – and reading obvious religious books – but not thinking at all about any religious vocation or order (anyhow you must go, not by books, but by a director)'

This may be related to a comment by Catherine in her letter to Newman of 28th July that Izy 'thinks he has been getting notions from books, in a theoretical way, and that he does not understand what he talks about – and is not aware that he does not understand.'

William wrote to Newman again on the 25th.

> 'I fear my letter of the 20th must have seemed to you – as indeed it was a hasty and over passionate one – you will I am sure make some allowance for a cry of pain – '

He continued saying that he did not want to 'press Paternal authority' and asking Newman to guide Eddy.

Newman also wrote to Catherine on 24th July, repeating the points which he had made in his letter to Eddy, written the same day and admitting that he did not know what to advise. It is interesting that Newman opened his letter of 26th July to Edmund as "My Dear Child", instead of his usual "My Dear Eddy", perhaps deliberately emphasising the boy's immaturity. It was a long letter pointing out that Eddy should have discussed his possible vocation with him rather than with Suffield.

Catherine's letter of 26th July to Newman throws an interesting light on William:

> 'Your letter to William came most opportunely yesterday evening, when all the young folks had gone on an expedition to Dartmouth and Dartington, – and William and I were alone. He showed me the letter and talked about it saying it was excellent common sense as well as most kind. But he sighed a great deal as we walked about the garden together, – and said much which there is no use in repeating, but which made me feel what a gulf there is between us. It is always extraordinary to me (seeing what excellent sense and judgement he has on most subjects) that in talking of Catholic matters, he does talk such nonsense. Such as "there can be nothing in the system of spiritual direction unless every director is infallible", as if one ought never go to a doctor unless the doctor is infallible. But I always comfort myself he talks more unreasonably to me, than he would to anyone else, – or even than he thinks. It is more by relief from the pain which oppresses him (just as one groans or cries out) and wives are meant to be beasts of burden. So I do not in the least mean to complain. But I do wish he could see things differently.'

This letter may well describe the first occasion on which William and Catherine had talked together over Eddy's problems.

Eddy was beginning to realise the trouble he had caused and asked Newman to suggest suitable penances. Newman's lengthy

Chapter 6: A Sacred Duty to Doubt

reply concluded that Eddy would be a better priest if he had a university education first.

William explained his concern in a long letter to Newman on 5th August saying that his position was different from that of a father whose son has no interest in the family business:

> 'It is not that Eddy does not feel an interest in the things that interest me, and that he takes to something else on that account – it is because of the great interest he takes in these things, and the very remarkable faculties he seems to have for seeing his way into the ideas on which the knowledge of such things hangs, and for developing them ...... and because in all these things I find his mind exactly in unison with my own .... that I am specially disappointed at seeing him resolve, on grounds which appear to me unreal to crush and throw away all these powers which he possesses and to me so interesting a degree......and give himself wholly to views which sever wholly the threads of sympathy between me and him in relation to all the deep questions of life -'

Another long letter from William on 10th August concluded:

> 'So far as the Oxford plan is concerned all I can ask of Eddy is that he should go through with the fair intention of working for honours in Mathematics and Natural Sciences – which will form the staple of his profession if he becomes an engineer, and for which he has great – very great – natural talents.'

Eddy never did go to University as he developed the family complaint, tuberculosis, and spent the next two winters at Menton to help his recovery.

This episode may seem to have been a storm in a teacup but, to those involved, it was a major emotional crisis. With hindsight, one may feel that this crisis need never have arisen had Edmund and Catherine been more open with both William and Newman. The latter could have taken a much stronger line – wait – in his first discussion with Eddy. William over-reacted to a sixteen year old's enthusiasm and there seems justice in Catherine's view that her husband talked 'nonsense', particularly to her, on Catholic matters; later she was to call it 'crankiness', a view given some support by Henry Brunel's journal entry. However, in exposing their deepest feelings in this matter, all concerned revealed much of their character and feelings. While some of Newman's letters appear to indicate that Edmund was immature, it should be realised that he was assisting his father to a considerable extent in the design and building of their house at Chelston Cross, discussed in the next chapter.

Father Suffield, who had caused the trouble, left the Roman Catholic Church in 1870 and became a Deist!

There was another family tragedy in 1868 when Arthur, too, developed tuberculosis and died on 12th April. While he was very ill, Catherine had written to Newman:

> 'I thank God every year more and more, that we have you for a

friend. It is curious to me to see that – although my children are all so different, yet there is something in your writings which fits into their minds in a way that no other serious reading does. I read your books over and over again to Arthur, now that he is ill, and he is never tired of hearing them ... I cannot help feeling that ... this quiet time at home, for thought and attention to his religious duties, will have been of great advantage to him.' (20th February 1868)

6.6 Copyright and Dedication

When Newman left the Anglican church, he disposed of the copyrights and existing stocks of his Protestant books to various friends, including William Froude. Froude received nearly 3000 copies of nine works for which Newman hoped to receive £400 (ref 6.5). From time to time, Froude's letters enclosed royalty payments to Newman.

In 1867, Newman agreed to republish some of these early works and had to ask Froude's permission as the copyright holder. Then, at the end of the year Newman asked Froude to transfer the copyright to Copeland. Over the years, Froude seems to have been involved in a fair amount of work with no profit and surprisingly little thanks.

In 1871, Newman decided to republish his early essays for the "British Critic" and wished to dedicate the volume to William, first asking Catherine if she thought he would agree. This led to a three way correspondence which sheds further light on the Froudes. William, in a long reply, referred to Newman's 'persistent affectionate regard' for him and

> 'on that account and in remembrance of the eager and almost passionate interest with which we used to look out for them and delight in them when they first appeared. What you have proposed would give me the greatest pleasure.'

He continued:

> 'And yet there is something I ought to say, I am always in fear lest I should be somehow sailing under false colours in your sight; I do not mean deliberately and with pretence but by allowing you somehow through your affection for me to think more of me and better of me as if somehow you were supposing that were you to probe me to the bottom you would find more solid ground for approbation, a more real root of what you would regard as "Faith" than a thorough search would in fact disclose, and to let you dedicate your book to me would be in some degree to take credit for possessing what does not belong to me. '

Catherine, writing on 1st August, made a very important point concerning her husband's way of thinking, saying that William 'is under a mistake in thinking that you do not understand or appreciate the degree to which he has diverged, not only from your views, but from your principles of arriving at views'

> 'I believe from what you have said when we have talked on the subjects, that you do understand him certainly. It seems to me that William is utterly removed from the common run of sceptics; and his mistakes appear to me to proceed in great measure from crankiness, and a sort of over-scrupulousness. But, indeed, whatever they are, you know the worst of them, that I am convinced, ...'

Catherine then tackled the wording of Newman's tribute – 'amid many trials of friendship'. She said

> '... in whatever trials he had suffered (viz: from our leaving him) you had done all you could to soften them. I know he thinks that – he has always said it.'

William agreed to accept the dedication, (the final wording appears in Annex 6.1), in his turn paying tribute to Newman's influence on him:

> 'Indeed throughout it is I who have been the debtor, and even whenever and wherever I have in opinion or mode of thought been not with you but against you, and I suppose against Hurrell also (though his probable mental course had he lived to these days has seemed to me more problematical) it has seemed somehow I have owed the power of thinking and of judging and above all the sense of responsibility to something which you and he taught me. there never has been a time when if I have been amongst those whose judgement I have thought well of or whose opinions I have been inclined to agree with, however at variance with yours, it has not added to the readiness with which I have been listened to and served in a sense, to give weight to anything I have had to say, if I have casually mentioned that I have been your pupil and was still counted among your friends.'

It is of interest that William saw his brother, Hurrell, and his employer, Brunel, as his principal guides in philosophy. He wrote to Newman on 24th December 1859: '...the peculiar manner in which the development of principles first cultivated in my mind by Hurrell was assisted and disentangled by my interviews with Brunel – a man of singular grasp of thought, and truthfulness and honesty of purpose, and whose views have often seemed to me to be most remarkably supplemental to, or explanatory of Hurrell's.'

6.7   Getting Old

During their long correspondence, William and Newman had sometimes discussed the passing years, for example, Newman wrote to William on 2nd January 1864 with New Year wishes – 'A new year is an awful thing at all times – but as one gets on in life, too solemn a thought almost for words. I recollect how I was oppressed when I was advancing to my lesser climacteric (age 49); and I am now close on my greater!' (age 63).

Then on 20th January 1870 William wrote 'I feel myself growing old and stiff – I am 60 next November. This drew a reply;

> 'I smile when you talk of your getting old and stiff, when I am 10 years older than you...'

References

6.1 The letters between Newman and Mr and Mrs Froude are all held in the Oratory, Edgbaston and are quoted with permission. Since they are dated in the text, they are not individually referenced. Some, but not all, are also quoted in (2).

6.2 G. H. Harper. *Cardinal Newman and William Froude – a Correspondence*. John Hopkins University, Baltimore, 1933.

6.3 R. E. Froude was known as Eddy to friends and family. The naval architecture profession usually refers to him as 'R. E'. Froude. (His father is usually referred to as 'William Froude'). I will describe him as Edmund to make it clear which Froude is being spoken of.

6.4 Henry Brunel's private Journal. Brunel Collection, Bristol University.

6.5 This section is based on the following letters:
11th March 1853
25th January 1867
26th November 1867
19th July 1871

Annex 6.1

To
William Froude, Esq.
To you, my dear William, I dedicate these miscellaneous compositions, old and new, as to a true friend, dear to me in your own person and in your family, and in the special claim which your brother Hurrell has upon my memory;– as one who among unusual trials of friendship, has always been fair to me, never unkind;– as one, who has followed the long course of controversy, of which these Volumes are a result and record, with a large sympathy for those engaged in it, and a deep sense of the responsibilities of religious inquiry and the sacredness of religious truth.

Whatever your judgement may be of portions of their contents, which are not always in agreement with each other, you will, I know, give them a ready welcome, when offered to your acceptance as the expression, such as it is, of the author's wish, in the best way he can, of connecting his name with yours.
I am, my dear William Froude,
Most affectionately yours,
August 1, 1871 John Henry Newman

Chapter 7
Paignton, Chelston Cross and the Brunels

7.1   1857–67

The decade following 1857 was a complete contrast to the preceding, relatively idle ten years. As well as his great achievements in the study of rolling and the early tests with the models SWAN and RAVEN which led to his work on resistance, William Froude was active in many professional areas, built his great house at Chelston Cross while his relationship with the Brunel family became much closer.*

Isambard Kingdom Brunel had become a neighbour of William's when he purchased land at Watcombe from 1847 onwards, near Torquay, and, with later purchases, it grew to an estate of 136 acres with a beautiful view over the bay. By the time of his death it was a magnificent estate laid out by Isambard himself. From 1849 onwards he and his wife rented a villa at Watcombe, now Watcombe Lodge, which was permanently staffed. His wife, Mary, lived there much of the time and was active in local life and charitable works. Isambard was very fond of the place and visited whenever work permitted and he, too, was involved in local affairs. He supported the local church, perhaps in too domineering a manner, and looked after the men working on his estate, arranging a special excursion to London for them during the Great Exhibition.

After his death, Mary spent several months each year in the Torquay area, even as late as 1876. It is pleasant to record that much of Brunel's great park will be preserved as a memorial to him and the other engineers of the Torbay area, largely due to the efforts of G. D. C. Tudor (ref 7.1).

The year 1859 was tragic and difficult. William's father died on 23rd February and Isambard on 15th September. In addition, William was upset by the conversion to Catholicism of his eldest son, Hurrell in December, followed by that of Eliza.

7.2   Paignton

Following the Archdeacon's death, the Froudes had to leave the parsonage at Dartington and they took a lease (from J. P.

---

*   Since several members of both the Froude and Brunel families appear prominently in this chapter, I will use Christian names alone after the first mention. These are I. K. Brunel, Isambard (1806–1859) and his wife Mary, nee Horsley (1814–1881), together with their sons Isambard (Jnr 1837–1902) and Henry (1842–1903).

Belfield) of 'Elmsleigh', to the south east of the junction of Fisher Street and Elmsleigh Road in Paignton. It was a large house, since demolished, standing in well wooded grounds and William soon equipped it with workshops and a small test tank in the attic. A large pond in the garden was also used for experiments. Though the Froudes saw it as only a temporary home until they built the house they wanted, they seem to have been very happy at Elmsleigh. Henry Brunel was a frequent visitor, complaining only of a plague of beetles so that the legs of his bed had to stand in pans of water.

7.3   Henry Brunel

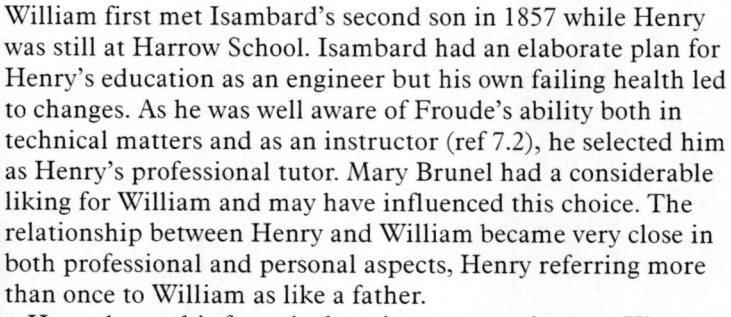

William first met Isambard's second son in 1857 while Henry was still at Harrow School. Isambard had an elaborate plan for Henry's education as an engineer but his own failing health led to changes. As he was well aware of Froude's ability both in technical matters and as an instructor (ref 7.2), he selected him as Henry's professional tutor. Mary Brunel had a considerable liking for William and may have influenced this choice. The relationship between Henry and William became very close in both professional and personal aspects, Henry referring more than once to William as like a father.

Henry began his formal education as an engineer at Kings College, London, in 1859, remaining until June 1861. He was always prone to grumble and found much of the teaching dull; at least in comparison with that of his father. He did appreciate the chemistry lecturer and some of the mathematics. There are a number of his letters from 1861 to the Rev. L. W. Owen, whose son wished to become an engineer, which shed an interesting light on engineering education of the day. Henry began by explaining the difference between Civil and Mechanical engineers. The latter dealing with the construction of engines and machines and the professional head of the company was usually the owner with only subordinate posts for others. The highest branch, Constructive Engineer, should possess a sound acquaintance with maths and mechanical principles. The usual way of starting as a civil engineer was to be apprenticed to a civil engineer with an undertaking that those of ability, 'regularity and business habits' should be given a permanent post on qualifying.

A mechanical engineer would charge apprentices a premium of about £200 while that for a civil engineer would be £500–600. To this should be added £2 per week for lodgings and, in the case of the civil engineer, £80 pa for advanced tutorials.

Whilst at College Henry took a keen interest in the building of HMS WARRIOR, the first iron hulled, armoured battleship, second only in size at that date to his father's GREAT

Figure 7.1 – Henry, second son of I. K. Brunel.

Chapter 7: Paignton, Chelston Cross and the Brunels

EASTERN. His contact seems to have been James Ash, brought in as naval architect by Mare in 1846, whose company later became the Thames Iron Works. Froude visited her in December 1860 but did not attend the launch despite strong persuasion from Henry who had been given tickets by Ash;

> 'I think the launch will be a very awful thing and I am afraid that my putting before you the possibilities of a catastrophe will not make you more anxious to come.'

However, Henry wrote on 29th December, 'the launch went off very well' though it was so cold that the grease froze and WARRIOR had to be pulled off the slip by tugs.

In January 1861, Henry wrote to Froude's oldest son, Hurrell, saying that she had 'a good deal excess of stability'. It is not at all clear what he meant; without engines, her stability would have been quite poor but as completed she was fully satisfactory and would probably be seen as adequate today. Froude visited the ship in February and March. Another of Henry's letters in March says

> 'Mr. Froude has been going into the calculation of the period of rolling of the WARRIOR and I have been helping him to the best of my ability. He has therefore been running up to Town a great deal. It is a very laborious job.'

None of this investigation has come to light and it can only be presumed that it was carried out at Ash's suggestion.

On 5th May there is a cryptic note, 'What about the WARRIOR. I don't know what is to be done about the horrid thing. Shall I go down in the dead of night and burn it, or shall we give up the job?' It seems that the work was abandoned.

In early 1861, Henry and Edmund Froude began to plan a very extensive model railway in the grounds of Elmsleigh. It was, of course, broad gauge with the 8 feet of the full scale represented by 4 inches in the model. The size was impressive; letters refer to platforms 6 feet long, a viaduct whose five middle spans were 30 inches, six at 25 inches and 8 more at 20 inches span. There was an embankment requiring 100 cubic feet of soil – '3 days work'. Stations and villages were given imaginary names such as Potaggerdoin, Appleford, Hedgley and 'histories' were written for them.

Edmund was 15 and Henry 19 and they took this work very seriously. As befitting the sons of such distinguished railway engineers, they planned the documentation for 'Parliamentary approval', Henry sneaking downstairs at Duke Street at dead of night to discover the current fashion for presenting the drawings. Unfortunately, their plans were over ambitious and the railway was never finished. Henry, with some help from William, also restored the timber viaduct over a road in the Watcombe estate which Isambard had built in 1852.

In 1862 Henry became a premium apprentice at Sir W. G. Armstrong & Co, Elswick, probably the most advanced engineering company of the day. William Armstrong was a close friend of the Brunels and Henry was allowed considerable latitude in holidays and other absences – but not every apprentice returns from holiday with cream for Lady Armstrong. Henry dined frequently with the Armstrongs and with the Nobles, later Chairman of the company, and also with his supervisor, Percy Westmacott, whom he described as a very agreeable gentleman. When his mother visited him, they would both stay with the Armstrongs.

The best description of his daily routine comes in a letter to Hurrell Froude (3rd December 1861) which, summarised, reads:

> 'At five and six am during the winter, he was reduced to abject misery by the alarum and rushing out of bed... Then he gets dressed, gets breakfast ready, coffee and milk in a tin bottle which goes in the pocket, a sandwich of cold meat goes into a box and walks to work.'

There was a half hour break for breakfast at 0830 and another one and a half hour break at lunchtime when he returned to his lodgings for dinner.

As a premium apprentice, he began in the pattern shop for three months with a 'nice, superior class of carpenter' which was to be followed by similar periods in the fitting shop, foundry and erecting shop, after which he would go to an outside job, returning to Newcastle for nine months in the drawing office. After the day's work was over, Henry returned to his lodgings for tea followed by letter writing till bedtime at 10. Though he grumbled – as usual – at the time, Henry was later to commend this apprenticeship, notably in a letter to Metford of 23rd November 1875. He learnt to appreciate materials, to use tools and the value of the different branches of labour.

Like many young men away from home he was inclined to complain in his journal and letters but these must be treated with caution. He was homesick and had all the usual problems of students; his landlady was 'a horrid old woman', his favourite wine ran out and funds ran low and his clothes wore out, all summed up before returning from one holiday as 'bloody Newcastle', though once he had settled down, this was amended by 'it isn't so bad after all.' On another occasion (June 1861) he referred to the pleasure of a few days in London with Edmund before returning to the 'dissolute wilderness' (Newcastle).

In fact he seems to have lodged very comfortably with Mrs. Brooks, paying her 15 shillings a week, though with many 'extras'. This rate was negotiated from an asking price of 16/-, Henry pointing out that the standard rate was 10/6. His wages were 3/6 per week but he had an allowance of £3 per week from his mother in addition.

He also complained that the work given to him was both boring and arduous and his supervisor was unimaginative. It must be remembered that he had been his father's pupil in

Chapter 7: Paignton, Chelston Cross and the Brunels

engineering since the age of seven and most teachers would seem unimaginative in comparison. Henry was given a letter press by William, to which we owe the copies of his letters, and he and Eddy with some help from William were full of schemes for improving it.

Armstrong allowed Henry to make frequent visits to the GREAT EASTERN and in June 1861 he and William accepted an invitation to voyage to Canada in that ship, meeting at Liverpool as William had been visiting the Speddings in Cumberland. The GREAT EASTERN was carrying 2144 officers and men of the 30th and 60th regiments, together with 6 Armstrong guns and 122 horses to defend Canada against Fenian raiders based in the USA. William and Henry were made welcome; Captain Kennedy was 'very civil' and the Chief Engineer, Rorison, 'most agreeable' while the officers were 'as a rule gentlemanly and content with their quarters but one or two exceptions.' The visitors were given single cabins measuring 8'6 x 6'6 on the ship's side, on the middle deck in the Second Saloon just abaft the paddle engine room.

Writing to his brother, Isambard (Jnr), Henry wrote that 'everyone was delighted with the ship,...eating is very fair and the steward's arrangements very good.' They left Liverpool on 27th June, passing down the Irish sea, rounding Cape Clear (Eire) and crossing to Cape Race. They kept detailed records of the ship's performance which may be summarised: the daily run was about 320–340 miles at a speed of about 13 knots, corresponding to 10 rpm on the paddles and 37 on the screw, and burning about 340 tons of coal a day. There was dense fog on the third day which persisted for the rest of the voyage while on 2nd July they met icebergs, one being 500 feet long and 180 feet out of the water, said to be as 'large as the rocks in Hope's Nose, Torbay'. William was a bit disappointed with the speed but blamed it on the coal; it is also notable to modern eyes that they did not slow much for fog or ice.

They also measured wind speed and pitch and roll but the calm which brought the fog meant that the ship did not move much and even Henry's sensitive stomach was not 'discomposed'. The maximum roll was about 3.5 deg whilst pitch at the stern was measured at 10 feet maximum. The passengers were well content by the voyage and had no objection to the troops whose two bands played for the passengers from time to time. The horses were on the upper deck aft and obstructed access a little. Henry thought it was as quick a passage to Quebec as any made, the engines worked well with not even a hot bearing. William and Henry left the ship on arrival to visit the Niagara Falls, Toronto Observatory, where they studied wind gauges, then to Montreal and Quebec, dining with the Governor General.

Henry, then and always, was inclined to leave things behind, writing to Edmund in January 1861 that he had left umbrella, silver pencil case, drawings and his gold toothpick behind at Elmsleigh. The toothpick was lost again in the GREAT EASTERN and he made do with the point of his compasses but the toothpick was found when the ship unloaded. It was a family failing as his brother usually left things behind whilst William – and Catherine – too, were forgetful.

In 1862 Froude and Henry worked on an indicator for use with the paddle engines of the GREAT EASTERN. The design work began at Paignton but Henry completed the detail drawings in his digs at Newcastle, a task which he found boring as he wrote to Eddy, 'We must be bored in this wicked world or things won't get done.' There were many problems but Henry was determined to succeed 'care, patience, perseverance and ability' would see him through. By early summer Henry was back at Paignton and working long hours. The recorder was built by G. D. Kittoe, formerly with Isambard, but last minute problems with springs forced a change to a design using a hastily ordered selection of rubber bands. It was finished in time for it to be installed by 30th June, in time for GREAT EASTERN's next voyage, by Rorison, the Chief Engineer, who was delighted with it. This was probably the only dynamometer made for paddle machinery.

7.4   Private Affairs

Isambard had established a family tradition of amateur theatricals at Christmas and by 1861 Henry had re-established the practise at Elmsleigh, presumably because it was bigger than Watcombe Lodge. His Aunt, Sophy Horsley, provided the musical accompaniment whilst Metford supplied fireworks which played an essential part. Their party piece was 'The Regular Fix' and they also performed 'To Oblige Benson'. The preliminary plans were agreed by letter from Newcastle and there would be a week's rehearsals followed by two or three performances, the first for family and close friends, the next for servants and locals. Mary's dislike of Catholics led her to disapprove of her son's increasing friendship with the younger Froudes and William, too, became upset by the disruption caused by these performances with curtains, chairs and other items removed from all over the house for scenery. Metford was a long standing friend of William's and letters survive in which they discussed, at length, whether the century ended at the end of 1899 or 1900.

By this time Henry was in love with Izy. There is a charming and, to modern readers, surprising, passage, four pages long, in his journal for 1863 in which he debates the ethics of kissing Izy 'goodnight'. His conclusion was to the effect that it was a bit naughty but rather nice and they decided not to tell their parents who might not understand. (He was 21 at the time and she was 23) Their affair came to an end when it became clear that Mary would never agree to her son marrying a Catholic. Henry was also involved with a married lady, referred to as 'Polly', and this led to a scandal in 1865 as a result of which William forbad further 'intimacy' with his daughter though Henry remained on good terms with the Froudes, corresponding regularly with Izy.

Two letters from this difficult time illustrate the depth of feeling which Henry had for William Froude. In the first, of 29th July 1865, he wrote

> 'Most of my pleasant reminiscences are of your house and in connection with Izy'.

On 10th August he said

> '...you are the only friend I ever had and almost every bit of good in me is due to my acquaintance with you. You know how I love you – please let me feel I have not hurt you.'

By 1861 William and Henry began discussing their ideas for a flying machine and when, in October 1861, William went to Newcastle for a few days Henry suggested that they go down 'a coal mine as the flying machine will go higher than anyone has been, they ought to go as low as possible first.' They were to develop the idea of flying until William's death; indeed his last paper may be seen as a contribution to their dream, for they always realised that available machinery and materials were too heavy even though one letter from Henry suggests a gas turbine or Hero engine. Beauchamp Tower was a fellow apprentice at Newcastle and his ideas on lightweight steam engines were drawn into the schemes for flying machines.

7.5   Yachting

William's interest in yachting remained keen though details of his activities are hard to find. The 'Exeter Flying Post' of 15th August 1844 refers to the loss of a boat belonging to a son of Archdeacon Froude while on passage from Exmouth to Dartmouth with the death of the young man crewing it. In a letter of March 1861, Henry refers to fitting a new mast in the Gannet but without detail and in October 1862 the Gannet was raced in the big cutter event at the Torbay Regatta.

In 1862 William was one of nine Stewards of the Royal Torbay Regatta though, surprisingly, he missed the Ball even though Catherine and Izy attended. By 1872 William had joined the Royal Dart Yacht Club of which his father in law Arthur Holdsworth was a founder member. Bidder the civil engineer was also a member.

In 1863, he began the design of a new yacht which seems to

have been of unconventional form since Henry wrote that William was well known at the quay for queer looking things, including yachts. The keel was laid in April even though the lines were still being argued about between William, Henry and a friend, Gray, who was preparing the drawings, as late as August. Beauchamp Tower was involved in the building.

At least two six foot models made of tin were tested, sometimes rigged, and it seems almost certain that one of these became the RAVEN, used in comparison with the SWAN in the experiments described in Chapter 9. In October 1863 there were some capsize tests, Henry noting that he went out in the large boat so that he could come back if he was sick. There is no record of this yacht being completed. Henry's letter of 27th January 1864 reads

> 'The yacht experiments which were first going unfavourably are looking up as they have succeeded in getting rid of the effect of some oscillations in the machine.'

By 1866 there are continuing references to the Gem, which, at first was borrowed or hired but by 1872, Hunt's Universal Yacht Register lists William Froude as the owner of the Gem of 23 tons (probably a ketch), which he retained till his death (ref 7.3).

7.6 Family Problems

There were continuing family problems which caused much distress to William. As discussed, Eddy became a Catholic in 1863, which William saw as incompatible with a scientific career, (Chapter 6) and the following year announced his intention to become a priest. Even Henry, who was far less concerned by religious matters, was upset writing to William on 22 July that he was on the point of crying and that with Eddy giving up the profession he was losing the one he looked forward to as having as a good friend. Henry continued with an important passage:

> 'I owe it to you of how little use I should have been if you had not taken me up – and been my father not withstanding the many annoyances I have caused you and how glad I shall be if all the good you have done me should be a comfort to you as I hope it may be and I wish I could put into words how much this has drawn me closer to you.'

On the same day, Henry wrote to Izy:

> '... though I lose a good friend and as I had hoped a companion in my profession I can only think and grieve for the great and good man who has been so much to me.'

Also that year the youngest child, Mary, developed tuberculosis and died in her father's arms in harrowing circumstances.

There is also an intriguing letter from Henry in 1866 in which he suggests that the accuracy of speed trials could be improved by fitting a dynamometer to the shaft 'on the hydraulic plan', possibly foreshadowing Froude's later dynamometer.

In 1866 Henry was working as assistant to Sir John Hawkshaw, builder of the Severn Tunnel, and now part of a syndicate planning a Channel Tunnel. Henry's job was to lease sites in France for trial borings and to extend these borings under the sea, which had not been done before. He brought in the Froudes, William, Hurrell and Edmund, (and others) to help design the apparatus needed including navigational aids, anchors, a punch lead and a dead reckoning buoy. The big job was a coring machine, very similar to those used to this day; the prototype was built by Froude in his workshops, followed by a larger version made by Kittoe. This was tested in Torbay in September 1866.

7.7 The Torquay Water Supply

Torquay was supplied with water from Dartmoor through a cast iron main, installed in 1837, with a total drop of 370 feet in its 13 mile length. The main had an internal diameter of 10 inches for the first 8 miles, where some of the water was drawn off for another town, and 9 inches diameter for the remainder. By 1843 the flow was only half of that estimated using contemporary formulae. All sorts of theories were advanced to explain the loss of flow. The accepted view of the day was that any roughness inside the pipe, such as that due to rust, could not account for the restriction since it was believed that the roughness would be 'filled in' and made smooth by a stationary film of water, leaving unobstructed the flow in the greater part of the pipe. It followed that the blockage must be due to accumulation of sediment in the lower parts of the main.

By 1863, the situation had further deteriorated and William Froude was called in to advise. He devised an accurate pressure gauge which could be tapped into the main without emptying it and, using this gauge, the pressure head in the pipe was measured at frequent intervals along the length. The loss of head per mile was uniform, showing that there were no local obstructions and that incrusted rust must be the cause of the problem. It also showed that the accepted view of the friction of water over rough surfaces was erroneous and Froude presented a paper on the subject to the British Association in 1869, discussed in a later chapter.

A well known civil engineer, J. G. Appold FRS, suggested that a spring loaded scraper could be forced through the pipe by the water pressure. The Water Board were reluctant to accept his idea at first, fearing that the scraper might stick and prove difficult to locate, let alone remove. However, they were persuaded by Froude to carry out a trial over a one mile section which proved very successful, reducing the loss of head from 21 feet to 7 feet which suggested about a 75% improvement in flow when the whole pipe was scraped. It was also found that the action of the scraper was so noisy that its position was always known so there would, at least, be no difficulty in finding it should it stick. Sadly, Mr. Appold died in 1865, the day before the scraper was successfully tried (fig 7.2).

Figure 7.2 – Three scrapers designed to clear the Torquay water main. Left Appold's original design with two others by Froude.

There is an amusing description of the test by Henry
> '...scraping rust in the pipe under the turnpike road with a loud rumbling noise – ridiculous to passers by, nothing above ground but 6 or 8 gentlemen and 2 or 3 navvies walking at 2 mile per hour along a quiet road, bobbing heads up and down in silence, pointing to the road, running on a bit, stick to ground and listening.'

Henry notes that he was up at 4.30 or 5 am both days and that the machine stuck on stones but William 'contrived' to clear it.

There were some minor problems and a new, improved scraper was designed by Froude with Box of Easton and Amos who made it. A cup shaped device was also made to remove the many stones from the pipe before the scraper was used. After the first complete scraping the flow was only increased by 40% but after further runs the delivery was increased to 655 gallons per minute instead of the 317 before work commenced saving

£30,000 over 20 years. Some time later, a third scraper was designed by Froude, and probably made by him in his workshop, which remained in use until about 1960. All three scrapers are held by the Torquay Waterworks Museum in Exeter (ref 7.4).

Froude's contribution lay primarily in the measurement of pressure and in the analysis of the readings, showing the true cause of the restricted delivery. His mechanical skill then played a small part in the final success. The consequences in other directions were also significant; his work was interrupted from time to time between 1863 and 1866 whilst between January and June 1866 everything else seems to have stopped, delaying his SWAN and RAVEN experiments. On the other hand his new appreciation of friction on rough surfaces would play an essential part in his later work on ship resistance (Chapter 13).

7.8   Chelston Cross

"....rather eccentric" (English Heritage)

The Froudes took their first step towards a new house in 1863 when William obtained a long lease of land at Cockington, Torquay, from his brother-in-law, William Mallock, on a beautiful site overlooking the town. W. H. Mallock wrote that

> 'During the first sixty years of the growth of Torquay, the owners of Cockington had preserved their rural seclusion intact, having refused, during that long period, to permit the erection of more than two villas on their property. But somewhere about the year 1860, a solitary exception was made in favour of Mr. William Froude, my mother's eldest brother, to whom, by my paternal uncle, a lease was granted of a certain number of acres on the summit of what was then a wooded and absolutely rural hill.' (ref 7.5)

Funds for the building came, presumably, from his inheritance from his father, possibly augmented by a legacy from his aunt, Margaret, who died at Denbury in 1862.

It was a very big house; the main frontage is 156 feet with a depth of 60 feet. Even though the Froudes entertained often, they do not usually seem to have had large numbers of guests and it is not clear why the house was so big though there is a reference in a letter of Henry's to the need for space to dance. As a hotel (1993) there are 18 bedrooms in the main house. This house, Chelston Cross, was always intended to be both a comfortable home and to provide research facilities and workshops.

It would seem that many people, too many, had a hand in the design. Froude, himself, had strong views, though these changed frequently, Edmund contributed much and Henry was willing to give his views.

> 'A letter from Mr. Froude almost dampened all hope of interfering with his house plans any more.' (Diary; 20 Nov 1863)'
> 'Had a shabbily short letter from Mr. Froude enclosing his plan for his house which I don't like at all.' (Diary; 21 Nov 1863'

There was a later letter of 13th January 1866:

> 'Isambard (Jnr) and Henry Brunel fought viciously about the dormer and I think Froude watched said viciousness and doesn't allay it but rather stirs it up.'

Later, in October of the same year, Henry said he was 'aggravated stupidly at Froude's chaffing too much and didn't like it'

Henry was assistant to Hayter at Hawkshaw's from October 1863 and was too busy to be deeply involved (ref 7.6). The one person whose views are not recorded is Catherine and one can only presume that she made her essential requirements clear to her husband at an early stage.

Henry seems to have irritated Froude at times with unwanted advice. Writing to Catherine (ref 7.7) he said

> '...has been something clumsy about HB's manner of suggestions about the new house which has aggravated Froude and made him cross..'

Eddy's contribution is more difficult to assess. Early in 1864 William wrote to Newman asking if Eddy might return to school a little late as he was proving so valuable in planning and supervising the work.

> 'The more so as one of the Brunels is (I hope) to be here just at that time, and he has also been giving so much assistance about our plans he and Eddy will be of great help in the consultation about many points which remain to be decided.'

Later in that year came his decision to become a priest, soon abandoned, followed by the onset of tuberculosis and two winters in Menton to recover.

The original architect, E. Appleton, appears to have had an impossible task in reconciling the requirements of all concerned, particularly as there were many novel features. By November 1867 Froude wished to end the contract with Appleton and in a long letter, Henry pointed out some of the problems (ref 7.8). He said that Appleton was entitled to 5% of the contract price of £5,000, ie £250, divided into preparing the design, contract drawings, working drawings and supervision. Though his work on the first two was inadequate, he would be entitled to £125. The working drawings were poor and Henry suggested that he should be paid for the paper and actual draughtsman's time whilst supervision was inadequate, Henry suggested offering £175 but in a later letter suggested going up to £200 to avoid a battle in court.

Appleton apparently claimed £306-1-2 and in a later letter (ref 7.9) Henry, after talking to his brother, Isambard, a lawyer, revised his views. While they felt that Froude would probably

Chapter 7: Paignton, Chelston Cross and the Brunels

win an action in court, there could be no certainty as Appleton would certainly claim that the house would have been much cheaper if built to his drawings, without Froude's innumerable changes, and that the charges based on the estimate of £5,000 rather than the actual cost of about £8,000 were generous. They thought that Appleton could make a fair claim to £200 out of £270 and it was not worth the risk of going to court over £70 – 'pay him.'

There were many features which were novel for the day. There was a lift, hot and cold water in each room, warm fresh air heating with a trench in the hall through which air passed over a hot water pipe and a lot of attention was paid to ventilation. It would seem that they did not get the heating and ventilation right as there were many complaints later (Chapter 13). There was a gas boiler and a 'steam gas engine'. (Presumably water gas made by passing steam over coke to give a lethal mix of hydrogen and carbon monoxide)

One of the most spectacular features of the house is a flying staircase which starts from a balcony round the hall running up to a landing on the floor above (fig 7.3). It was built as a 'bow string girder' bridge (as Sydney Harbour), canted up. It is constructed entirely of wood, fastened together with trenails,

Figure 7.3 – The magnificent staircase attributed to Henry Brunel.

and is graceful, light and rigid. Happily, it has survived and may still be seen today as an example of high quality technology in wood. There is a very strong oral tradition, both at Torquay and at Haslar, that this staircase was designed by Henry Brunel but there are few references in his papers. There is just one reference to William, Edmund and Henry visiting Appleton to discuss 'our' flying staircase. Design is a word which means very

Figure 7.4 – Chelston Cross, Froude's house at Torquay.

different things to different people. It may well be that the concept was Henry's and that the Froudes did the details, rightly crediting Henry as the designer.

There is an amusing male note in a letter of Henry's concerning the washstands in the bedrooms –

'luxury in washing means liberty to splash, surfaces to be watertight and waterproof. Marble most common...'

Henry suggested that Torquay shops might be old fashioned in the style of wallpaper available and offered to find new styles but it seems that this offer was not taken up. He also offered advice on dados and presented a beer engine for the dining room. He was not happy with the room in the attic which was set aside for his use and offered through Izy to supply his own bed, an offer which was withdrawn as she thought it might cause offence.

Froude was clearly aware of the problems with the Torquay water supply and, originally, the gutters had a single downpipe leading to a large storage tank below the house. Presumably for aesthetic reasons, the open gutters ran through some of the gables, inside the bedrooms, (just visible in fig 7.4) an inconvenient arrangement since done away with. (After 1872) Froude also agreed with the council to be responsible for the roads in the vicinity, re-making them with camber to improve drainage.

The layout of the grounds also suffered from too many advisers. Mary Brunel was an enthusiastic gardener and pressed her views strongly, as a result of which the Watcombe Head Gardener, William Elston was engaged as was Veitch, as nurseryman. Henry, too, gave his views and may have been responsible for an artificial mound or 'Mountain' similar to that

Chapter 7: Paignton, Chelston Cross and the Brunels

at Watcombe.

Extensive workshops were provided in outbuildings which may easily be mistaken for stables and which are now the hotel staff quarters. As usual, William was proud of his workshops and, when lecturing to the Institution of Mechanical Engineers at Penzance in 1873, he invited members to visit Torquay on the way home and see them. The house had a big semi basement, 114 feet by 47 feet, now the ballroom, which William and his friends used for experiments and the assembly of large items of equipment. Henry used this space extensively and was a frequent visitor to the house. Another, later, user of the laboratory was a fellow apprentice of Henry's, Beauchamp Tower, who was developing a lightweight steam engine.

It was clear that a tank was needed for experiments and it was suggested that it should be in an outbuilding, attached to the house, 'which is to look like a domestic chapel'. This tank is now the hotel swimming pool and its present length is 27 feet. Other than tiling, it is likely that this is the original dimension as the exterior, little changed, would not permit a longer one. The exact purpose of the tank is not clear. It was certainly used for rolling work, see, for example, Reed's visit in 1868 and the GREAT EASTERN demonstration but it is far too short for any serious resistance work. There is a reference to an 'impulse apparatus' but it is not clear what is meant.

The house was finished in 1867, Henry, as usual, pushing the older William into action. To this end, William agreed to cut out tea breaks but spent the time saved playing with 'his repulsive

Figure 7.5 – The small tank built on to the house for rolling experiments and disguised as a chapel. It is now the hotel swimming pool.

dog Crack who would end in Hell'. There is a gap in Froude's notebook from April to July 1867 which must mark the delay to work caused by removal and setting up the new laboratory and workshops. Finally, Henry wrote to his friend Mary Noble (ref 7.10) that architecturally it was a great success and those who come to scoff remain to praise.

Chelston Cross was a happy home as well as a workplace. Harper (ref 7.11) lists the visitors as including Arnold, Jowett, Ruskin as well as Henry. Admiral Popoff, Imperial Russian Navy, and designer of circular ironclads was another. Froude was a member of the Athenaeum and 'by reason of his personal charm and intellectual distinction a popular figure in London Society.'

Pengelly, the geologist, was another visitor and his daughter (ref 7.12) has published some of the few letters by Froude which show signs of humour.

> 'Chelston Cross, 11.59pm (That is to say after the eleventh hour). Sir Thomas and Lady E. are coming to dine with us tomorrow. .....and if you can spare the time, and it isn't a bore to you so to spare it, it would give us and him pleasure if you would join us and him at dinner to-morrow (you see by this time it is today!) at 7.30pm. Don't hesitate to say no if you would rather not! but make up (in that case) some agreeable excuse, such as that the bottom had fallen out of Kent's Cavern, or that the earth contained in it is in apogee, or that you are in opposition, or something more pleasant even if less true ....' (Undated, but probably ca 1870)

Another letter refers to what William calls 'Foolish sham science prognostications.' continuing

> 'It is a curious psychological fact that people will listen in the name of science to any quack who talks sham science, but are disposed to think that such men as the Astronomer Royal and his establishment are ignorant or inattentive to their business.'

Pengelly was clearly a good friend as, together with Professor Tyndall, he proposed William Froude for a Fellowship of the Royal Society in 1869, Froude writing;

> 'This is the third time I have been balked of my purpose of calling on you, to thank you for the very great and effectual trouble you have taken on my behalf in reference to my F. R. S. ambition... So let me now thank you on paper at least. You are kind enough to make light of it; but I know what a tax it is on the time of a man who occupies his time fully to write a lot of letters, and this tax you have paid for me without stint.'

William then gently pointed out a minor inconsistency in Tyndall's theory of the effect of water vapour in the earth's atmosphere and its effect on temperature.

William Froude had now achieved professional recognition, financial security, a fine home and, with the exception of religious matters, a happy family life.

Chapter 7: Paignton, Chelston Cross and the Brunels

References

7.1 In February 1987 the 'Brunel in Devon' Trust (Now 'Brunel in Torbay') was established to study the Brunel family's involvement in the area and to preserve evidence of their achievements. The main focus has been on I. K. Brunel's Watcombe Park, which he laid out between 1847 and 1859 on the northern outskirts of Torquay. This is now registered as an Historic park and forms the nucleus of the Watcombe Park Conservation Area. In December 1992 a memorial column to Brunel 22 metres high was erected at the lower end of the Watcombe Valley, carved by Keith Barrett from a Wellingtonia blown down in the storm of January 1990. Brunel's achievement in landscape design is at last receiving recognition, due largely to the efforts of G. D. C. Tudor.

7.2 Much of this chapter is based on letters by Henry Brunel held in the Brunel Collection in the Bristol University Library. I am also grateful to Messrs G. D. C. Tudor and G. Maby for drawing my attention to key passages and to the latter for making his summaries of the letters available. Only a few key letters are separately referenced, dates in the text make the location of the source simple.

7.3 Much of the information on William Froude's yachting activities has been supplied by Ms Janet Cusack, Exeter University, from her thesis on yachting in Torbay.

7.4 W. F. White. Note on the scraping of the Torquay water main included in Froude's Collected Works, facing page 137.

7.5 W. H. Mallock. Memoirs of Life and Literature. London, 1920.

7.6 G. H. Harper. Cardinal Newman and William Froude - a Correspondence. John Hopkins University, Baltimore, 1933.

7.7 Mrs H. F. Julian. James Anthony Froude and His Brother, William Froude, FRS. Devonshire Association, 1921.

The Way of a Ship in the Midst of the Sea

Chapter 8
Ship Hydrodynamics, Background

8.1   Newton to Froude

It is convenient, and generally correct, to see Isaac Newton's 'Principia' of 1686 as the first attempt to produce a theoretical explanation of the motion of bodies through a fluid. In Book II, section VII he considered the fluid as consisting of a very large number of small particles which impacted on the body losing momentum themselves and transferring it to the body. The resultant forces could then be summed over the whole surface of the body, giving the total resistance to motion. Newton found, that if viscosity is ignored, as he did, the total resistance of a deeply submerged body, where there is no wave making, is zero, a finding now known as d'Alembert's paradox (ref 8.1).

In order to get what he saw as a sensible answer, Newton then considered the forces on the fore body only and was able to show that resistance varies as the square of the speed. He also believed that there was a unique shape for minimum resistance at all sizes and speeds. Newton's approach was copied and extended by virtually every hydrodynamicist for the next two centuries. This work was ingenious but fundamentally flawed and about as relevant as the phlogiston theory of combustion since viscosity was largely neglected. Wave making was not properly considered and the effect of the after body was ignored (ref 8.2).

The French naval architect, Bourguer, published his 'Traite de Navire' in 1746 which still used Newton's particle approach and concentrated on the shape of the bow and midship section. In England, Newton's work was developed by Sutherland in two books, published in 1711 and 1720.

Duhamel du Monceau published 'Elemens de l'Architecture Navale' in 1752 following the Newton-Bouger approach but with improved mathematics. This work appeared in English in Mungo's 'Shipbuilding and Navigation', 1754, and was much extended in a later edition in 1765. John Bernouilli, 1740, and Bossut, 1771, showed some of the fallacies in Newton's work but still omitted viscosity and implicitly directed attention to the fore body.

In 1775, d'Alembert, Condorcet and Bossut carried out some careful experiments on models in the grounds of the Ecole Militaire, Paris, primarily aimed at canal navigation. These experiments were thought to be of value to seagoing ships as well and have frequently been referred to with respect. They seem to have suggested that the effect of friction was negligible but did not show the importance of the bow wave and disproved more of Newton's numerical results.

In England, the Society for the Improvement of Naval Architecture sponsored a long series of model tests by Colonel Beaufoy in Greenland Dock, Deptford (ref 8.3). Between 1793 and 1798 he carried out some 1670 successful runs (not the 10,000 frequently quoted). The results were published in 1800 and in a much more complete form by his son in 1834.

Beaufoy followed a suggestion by the Earl of Stanhope, a famous naval innovator, dividing total resistance into two components:

Friction due to water rubbing along the surface of the ship.

The algebraic sum of the pressure over the fore and after bodies.

This was an enormous improvement on previous concepts of total resistance and, carefully defined, would be seen as valid today. (See Annex 8.1)

Dr. Tom Wright has shown that Beaufoy's measurements of frictional resistance were little different in value from those obtained much later by Froude. However, Beaufoy's graph has a much smaller slope with size and speed than Froude's and, in scaling, it is the difference between the values of friction for the model and the ship which matters. Beaufoy's results would not have given a good answer. On the other hand, Beaufoy's work was a great advance over the Newtonian approach and he did show that frictional resistance increased at a power of speed less than the square (about 1.7 to 1.8).

Beaufoy's work came to end with the collapse of the Society about 1800 and there is no evidence that he understood how to scale from model to ship. Froude's writings make it clear that he had seen Beaufoy's work and accepted much of it. There were many other careful model experimenters in the following years such as Palmer (1824), Robert Rawson and Fincham (1850), Napier and Scott Russell, the latter at least using Beaufoy's work on friction.

During the 1820s, the Admiralty was determined to improve the performance of sailing warships and carried out a series of full scale trials known as 'Experimental Sailing' extending over two decades. Special trials ships were commissioned from various designers. These trials were carried out with great care and show the enthusiasm of the Admiralty and the participants but little of importance was learnt. The results were too dependent on the sailing skill of their captain and on his ability to find optimum trim, rake of masts, adjustment of rigging etc. It was also clear that a ship which was clearly superior in one condition of wind would be inferior under different conditions of strength and direction (ref 8.4).

By the end of the sailing ship era there was no understanding of the nature of ship resistance or of the way it was influenced by the form of the ship. British Master Shipwrights, mostly well educated men, are frequently criticised for failing to use French

Science or even failing to be aware of it. Both charges seem dubious, English translations of French works were sold fairly soon after publication and there must have been purchasers. More probably they rejected the concentration on the fore end as experience would have taught them that the afterbody shape was at least as important. the area and shape of the midships section was seen as important by all as was the belief that there was a single optimum form.

8.2   Resistance and Power

All the work described above relates to sailing ships in which the action of wind on the sails overcomes the resistance to motion of the hull through the water and the designer's objective was to reduce that resistance. The problem with a steam ship and in particular with the propeller driven ship is much more complicated due to the interaction between the flow round the hull and that of the propeller as shown by Froude in his 1850 experiment on Lake Bassenthwaite, which had never been published (Chapter 3).

It became customary as was later formalised by the Froudes to define 'resistance' as the force opposing the motion of the bare hull of a ship or model when towed through water by external action. The power needed to overcome this resistance (R) at a speed V, later defined as the effective horsepower' (ehp) was
  ehp=R.V
This differed from the power put into the propeller shaft by the engines, 'shaft horse power' (shp) due to losses in the propeller and by the effects of interaction between the hull and propeller, discussed later(Chapter 13).

There are several physical actions, such as friction, other viscous effects and wave making which contribute to resistance but they all interact (see Annex 8.1). However, the overall problem is too complicated, even today, to tackle as a whole and Froude's greatest achievement was to break resistance down its two components which could be handled separately and used to scale the results of model tests to ship performance.

8.3   Steam Propulsion; its Development and Problems

During the 1820s the number of steamships grew steadily and by 1830 there were some 300 British commercial steamers, mostly quite small, and eleven Admiralty vessels which were considerably bigger than most. Most of these were ferries or tugs and little attention was paid to improving their performance or efficiency; it was enough that they worked at all.

By 1840 there were over 800 merchant steamers and 45 RN ships, some with considerable fighting capability, almost all powered by side lever engines, a modified beam engine, driving paddle wheels. The machinery was big and heavy occupying about 40% of the length of the warship. It was not possible to estimate the power required for a given speed, instead, the engine designer was given a weight, the desired height of its centre of gravity and a drawing of the engine room and told to install the 'best' engine he could.

The power of the engine was usually stated as 'Nominal Horse Power' (NHP) (ref 8.5), a fictional figure based on the area of the piston and on an assumed and usually incorrect piston speed which differed from the actual power by a factor of up to three. The NHP was a fair representation of the size of the engine and hence of its cost. Attempts were made to equate NHP/ton with speed and tonnage, without success. The 'indicator' which measured changing pressure inside the cylinder during a complete cycle gave a fair measure of the power inherent in the steam and, interpreted as 'indicated horse power' (ihp), had a real, if complicated, relationship with the engine's output. However, losses due to friction inside the engine and in driving essential auxiliaries such as the air and water pumps meant that the power at the end of the propeller shaft was much less than indicated. Trials of WARRIOR in 1861 suggest that the shaft horse power (shp) was about 75% of the indicated power. RATTLER's trials in 1845 are less easy to interpret but the output (shp) was perhaps only 65% of the power in the steam.

In the late 1830s, Petit Smith and Ericsson, working independently, developed screw propulsion and, though it was quickly appreciated that the propeller was more efficient than the paddle wheels and, particularly in warships, conducive to a better and safer layout, it was very difficult indeed to match geometry of the screw, diameter, pitch, area of the blades etc, to the characteristics of the hull and its machinery. There was no real understanding of how a propeller worked and, without such understanding, model tests could not be related to full size performance.

There is a well known story which illustrates this lack of understanding. Smith's trials launch, FRANCIS SMITH, which had a long propeller of one and a half complete turns, lost half its propeller on hitting wreckage and went faster as a result. Later, his demonstration ship ARCHIMEDES, had several different propellers before achieving her best results. She was tried by the Admiralty soon after completion and in a series of trials – races – beat, or at least equalled, the performance of the fastest paddle mail packets of the day, some of them with much more power. As a result of the enthusiastic report by Captain Chappell and Thomas Lloyd, later chief engineer of the Navy, it was decided to build a prototype screw warship. Ericsson, too, had to change the design of his propeller radically after early trials.

The design was, at first, the subject of squabbling between three brilliant engineers involved, I. K. Brunel, Smith and Lloyd, but eventually they achieved mutual respect and became good friends (ref 8.6). This initial disagreement has been much exaggerated; as Henry Brunel wrote in his journal whilst editing the biography of his father, Isambard (ref 8.7), 'Have read about three quarters of the admiralty letters from the tin box and as far as I have gone I have not found much proof of the obstruction Capt. Claxton told us of.'

RATTLER completed late in 1843 and, after some preliminary trials, was modified during the winter including the fitting of a thrust meter at Lloyds's suggestion. In 1844 there were 28 trials with propellers of different size and shape from several designers. It was realised that rotational speed, diameter, pitch, length along the shaft axis and number of blades could all affect performance. Even as late as 1870 the choice of propeller characteristics relied on judgement with a good slice of luck; often it was necessary to design and make new propellers and carry out repeat trials at considerable expense. Luckily, the screw propeller and the slow speed steam engine are very tolerant and efficiency does not fall off quickly on departure from optimum conditions.

The best propeller tried, designed by Smith and happily preserved in the Naval Museum, Portsmouth, was used on RATTLER during competitive trials with a paddle half sister, ALECTO, during 1845. The ships raced under steam, under sail and using both. Each towed the other and, finally, they were lashed stern to stern for a tug of war. In all these events the screw was the winner, though careful analysis shows that with the engines at the same power the superiority was not very great (ref 8.8). The screw machinery and its supports weighed less, could be arranged below the water line for protection against gunshot, enabled more guns to be mounted on the broadside and worked much better in rough water. These advantages convinced the Board of the Admiralty of the screw's superiority and a big programme of propeller driven ships was started before the trials were complete; indeed the tug of war may be seen as a public relations exercise.

Hull form, too, affected performance. Lloyd, as Chief Engineer of the Navy, arranged several trials, notably on DWARF, to study the effect of varying the stern lines on propeller performance. There was little scope for variation in the shape of wooden ships as their weakness in shear meant that buoyancy had to be closely equal to weight at each section along the length leading to very bluff bows and sterns, far removed from a hydrodynamic ideal (ref 8.9). Iron hulls, introduced in the 1840s, once Airy had solved the problem of compass correction for the Admiralty, made it possible to design with much finer ends.

Later the Admiralty Coefficient was defined as

$$\frac{\Delta^{2/3} V^3}{HorsePower} \qquad \Delta = displacement.$$

During the 1840s the 'Admiralty Coefficient' came into use as the ratio of the indicated horse power to the midships section area by the speed cubed. This coefficient (ref 8.10) was based on the primacy of the midships sectional area, following the Newtonian fallacy previously outlined. A value of the coefficient could be calculated from trials with an existing ship which could then be used to forecast the ihp needed for a new ship provided that hull form and machinery were reasonably similar. It was also only valid at fairly low speeds as at higher speeds, where wave making was important, the power required increased much more rapidly than with the cube of the speed.

The idea grew up that if sufficient data could be collected and analysed in the form of the Admiralty Coefficient (or similar), it would provide a rational and adequate base for estimating the power needed in new ships. This was pursued through various committees of the British Association for the Advancement of Science who asked ship owners to supply trial figures. These committees, listed below, had as members the most outstanding men of the day and their failure should not be denigrated though such failures are a measure of Froude's achievement.

British Association Committees

Title	Dates	Membership
Mercantile Steam Transport	1855–59	C. Atherton
Shipping Statistics	1857–58	J. Scott Russell C. Atherton C. Moorsom
Steam Ship Performance	1858–63	As above
Condensation and Analysis of Steam Ship Performance	1863–69	J. Scott Russell J. R. Napier W. J. M. Rankine
Resistance of Water Floating and Immersed Bodies	1863–66	As above with W. Froude
On the existing State of Knowledge on Stability, Propulsion and Sea Going Qualities	1868–70	C. W. Merrifield W. Froude W. J. M. Rankine D. Galton F. Galton

The first such committee was on the form of ships, set up in 1838 under J. Scott Russell to examine his wave line theory, discussed a little later. Atherton's Committee of 1857 made a small step forward, redefining the Admiralty Coefficient in term of Displacement to the power 0.66 instead of midships area. The biggest 'ship owner' was the Admiralty and, over the years, they supplied several volumes of very detailed trials data on all sorts of ships. The only fact not presented which a modern naval architect would look for was time since the last docking, a crude measure of the increased resistance due to fouling. In the very few cases in which it has been possible to check, the trials were within a few days of undocking. The general level of care in the conduct of these trials suggests that this was probably normal practice.

However neither at the time nor more recently using modern statistical methods has it been possible to deduce any real conclusions from the data. As listed above, the British Association tried again and again, in some cases handicapped by a determination to force the facts into agreement with the wave line theory. (Discussed later)

The 1863 'Resistance' Committee was tasked with carrying out model tests to study the laws of resistance. The experiments were conducted by Scott Russell and provided no useful information. It seems that he failed to run the models at 'corresponding speed' in which the model speed is to ship speed in the ratio of the square root of their lengths. Such an error is a little surprising; this law was put forward by the French Mathematician, Reech, in 1844 and was published in England by Bourne in 1852 and was certainly referred to by another British Association Committee in 1869. furthermore, William Froude, a member of this Committee, had explained the principles of 'corresponding speed' very clearly in his remarks on Scott Russell's own paper on rolling in 1863 (Chapter 4).

8.4 The Existing Knowledge Committee

'Appointed to report on the state of existing knowledge on the Stability, Propulsion and Sea Going Qualities of Ships and as to the application which it may be desirable to make to Her Majesty's Government on this subject.' – Report prepared by C. W. Merrifield.

It seems from Froude's 'explanations' quoted in Chapter 9, that the report was entirely Merrifield's work and was not discussed by the full Committee. The report is a thorough survey of published work and, in some of its details, foreshadows Froude's work. However Merrifield makes it clear that there was no reliable way of estimating the power required to drive a ship at a specified speed and also that there was no valid supporting theory.

Merrifield breaks resistance down into the following components:

   Direct head resistance
   Skin resistance (friction)
   Back pressure

together with small components due to 'capillarity' (surface tension) and viscosity. This breakdown is clumsy in that it mixes a phenomenological division (friction) with geometry (head resistance) and suggests that he did not understand the problem.

Throughout the report it is assumed that resistance will increase with the square of the speed even though it is stated several times that this is not in accordance with measured results. the Admiralty co-efficient, in both its forms, is described as convenient though its failings are recognised. Considerable space is devoted to Dupuy de Lome's paper to the French Academy in 1865 which includes Reech's proof that, in an inviscid fluid, resistance per unit displacement will be the same for similar forms if speeds are in the ratio of the square root of the lengths. Merrifield specifically mentions that most earlier model experimenters assumed that the model should be run at ship speed and that this work of Reech's showed the error but Merrifield failed to realise that this work made model testing feasible. Beaufoy's work is also described.

Merrifield quotes Rankine as saying of wave making that its associated resistance is 'insensible up to a given limit of speed' and that Scott Russell's wave line theory shows how to design a form which set such a limit above normal operating speeds. This is not quite so silly as it seems to a modern naval architect as ships were usually slow and wave making resistance was not a major problem and, in particular, its departure from a square law with speed was small. For example, wave making accounted for less than 30% of the total resistance of GREYHOUND at 11 knots and even WARRIOR probably the fastest ship of the day at 14 knots, had a similar percentage of wave making at full speed.

Merrifield proposed to tow a full size ship with the propeller removed and measure its full scale resistance. He recognised the difficulties of such a trial and proposed suitable measures to overcome them. He said that, if possible a second ship of a very different size and proportions should also be tried and that the rate at which it slowed when the tow was dropped should be measured. Merrifield's proposal seems at first sight to be very similar to Froude's trial with the GREYHOUND (Chapter 12) but is totally different in that Merrifield gave no indication of how the results would be used.

The section of the report dealing with stability is satisfactory but that dealing with rolling is muddled; though Froude's work is described, it is mixed with less reliable ideas by others.

This inability to forecast power requirements is nicely illustrated by the design of HMS WARRIOR by Isaac Watts, one of the most skilled naval architects of the day and his outstanding colleague, the Chief Engineer, Thomas Lloyd. They used the Admiralty Coefficient to estimate the power for thirteen and a half knots and selected the engines on this basis. On trial, she made 14 knots and though this error of 4% may seem trivial it corresponded to a 12% overestimate of power. Cost was roughly proportional to power so the engines were, at £74,400, 12% more expensive than necessary. The bigger engines would also burn more coal, increasing running costs throughout her life. Finally, the effect of radically different hull forms could not be studied using the Admiralty Coefficient.

8.5 Waveline Theory

From the late 1830s until about 1870 ideas on hull form were dominated, at least in the United Kingdom, by the waveline theory of John Scott Russell. In 1834 he had been asked to carry out model tests to explore the way in which fast canal vessels could use the 'wave of translation' to get high speed with little power or surface disturbance. Such waves are set up in shallow water when the speed is equal to the square root of the depth multiplied by the acceleration due to gravity. It is possible for a canal craft to ride on such a wave at minimum power.

Scott Russell realised that this phenomenon would not occur in deep water but it focused his thinking on wave generation. His views changed from time to time and were never clearly set out (ref 8.11) but he suggested that a long fore body with sinusoidal waterlines would accelerate the water gently, minimising wave generation and hence resistance. He paid less attention to the after body and seems to have thought this could be shorter than the entrance and, at least later, suggested a trochoidal shape. Merrifield's report noted that there was 'some unsettled ground in his theory relatively to the shape of the after body.'

This theory was wrong in principle but it led to forms with fine ends with buoyancy concentrated towards amidships; what would today be called a low 'prismatic' form (ref 8.12). Such forms work well when the speed in knots is about equal to the square root of the length in feet as were many of the ships of the day. It is likely that Scott Russell, who was very observant, noted that models with fine ends disturbed the water least. Lacking a clear understanding, one may say he was right for the wrong reasons. On the other hand, Scott Russell's wave line form for the GREAT EASTERN was, though he did not know it, optimised for about 24 knots instead of the 14 knots of her actual speed, confirmed many years later by Edmund Froude's model tests (Chapter 14). Luckily Scott Russell was a very practical designer and the ad hoc changes he made, filling out the ends to ease layout, actually improved her performance at her operating speed. Scott Russell did concentrate people's minds on wave resistance and, while his theory was misleading, he produced some good ships.

References

8.1 This section is based very largely on a PhD thesis by Dr T. Wright, Science Museum/CNAA of 1983 for which the author was external examiner.

8.2 D. K. Brown. The speed of sailing warships, 1793–1840; an examination of the evidence. Conference – Empires at peace and war. Portsmouth, 1988. Published, Service historique de la marine, Vincennes, 1990.

8.3 These experiments are discussed at length in (1) and summarised in: D. K. Brown. Before the Ironclad. Conway Maritime Press, London, 1990.

8.4 As 8.2.

8.5 Nominal horse power was defined as:

$$NHP = \frac{7 \times \text{area of piston} \times \text{equivalent piston speed}}{33,000}$$

but equivalent piston speed was also arbitrary: see (3). Some readers may note the similarity to the old RAC rating for cars.

8.6 As 8.3.

8.7 Brunel Collection, Bristol University library.

8.8 As 8.3.

8.9 As 8.2.

8.10 As 8.3. The Admiralty co-efficient was probably introduced by Thomas Lloyd. See also The Admiralty Coefficient Shipbuilding and Shipping Record, 3 January, 1924.

8.11 Dr T. Wright. As 8.1.

8.12 Prismatic coefficient is defined as:

$$\frac{\text{Immersed volume}}{\text{Length} \times \text{Mid Sec Area}}$$

and represents the longitudinal distribution of buoyancy. Low prismatic forms have fine ends and are full amidships.

Chapter 8: Ship Hydrodynamics, Background

Annex 8.1 – The Components of Resistance

The total resistance of a model or of a ship is associated with a number of physical effects such as pressure changes associated with wave making, friction over the hull surface and other viscous losses. These phenomena interact; for example, due to viscosity the water nearest the hull will be dragged along with it, the so-called boundary layer, which will alter its apparent shape as regards the generation of waves. In consequence, there can be no exact breakdown of the total resistance force into components but on the other hand, it is essential to subdivide the total since viscous forces and pressure forces obey different scaling laws from model to ship.

Fig 8.A1 shows how total resistance may be broken down by physical cause or by the way in which the forces act on the hull. The upper lines are modern, involving complicated experimental techniques which only became available after World War II. The total force can be divided into components perpendicular to the hull (normal in the figure) and along the hull surface (shearing) without distinguishing how these forces arise. The next breakdown is into 'wave making', measured by exact measurement of the wave pattern generated by the model and 'viscous' deduced from a velocity survey of the wake. The measurements and their analysis are very complex and this process is a research tool, used only rarely.

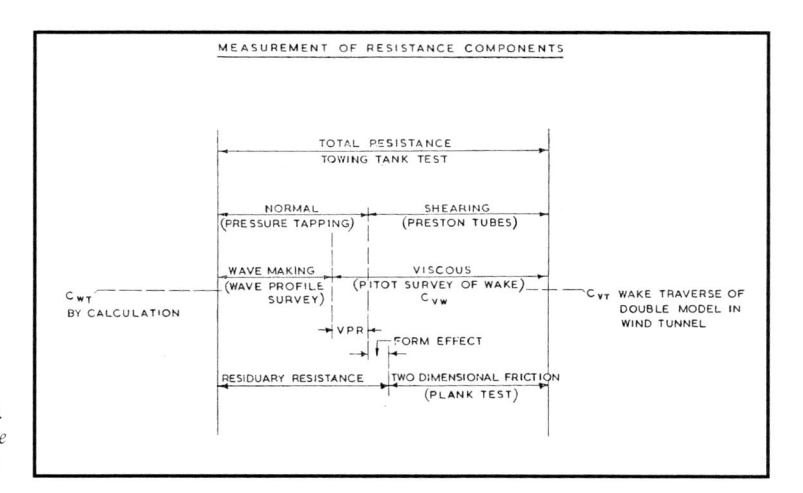

Figure 8.A1 – Different ways in which the total resistance of a ship may be subdivided. Any such division is an approximation since all components interact, one with the other.

This latter division is compared with Froude's simple frictional and 'residuary' breakdown at the bottom line. Froude assumed that the friction over the hull was the same as that over a flat surface of the same length, area and surface finish. this is not quite correct and most modern ship tanks allow a correction for form effect which is usually small.

The viscous pressure resistance is a little more complicated. As the bow moves into undisturbed water it changes the flow and causes a rise in pressure at the bow. If there was no viscous effects, this increase of pressure at the bow would be exactly balanced by a rise in pressure at the stern as the water returns to the undisturbed state. Viscosity causes a loss of velocity over the hull reducing the pressure rise at the stern leaving a net pressure difference acting as a force to oppose the motion. This viscous pressure resistance can be large in bluff forms.

Froude's description of 'eddy' resistance, outlined in Chapter 13, makes it clear that he understood the nature of viscous effects quite well but believed that the error in including this component with pressure forces as 'residuary' resistance was small. For most ships, this assumption of Froude's works quite well and the small error is corrected for by a factor of experience based on trial results. Very bluff forms, such as modern supertankers, need special treatment and the form of wooden steam battleships would have led to serious difficulties. Indeed, it seems likely that Froude would not have been able to justify his approach much before the 1860s when natural evolution had led to finer forms and these had been made possible by iron hulls.

Chapter 9
SWAN and RAVEN Experiments, 1865-1867

9.1 Early Days

Froude's interest in the performance of ships and of yachts led to a number of early experiments with models such as those on Lake Bassenthwaite and his tests with a model of his projected yacht on the Dart in 1863 (Chapter 3). These tests with yacht forms continued and merged almost imperceptibly into a more fundamental study of the laws of resistance and the interpretation of model tests.

Froude's notebook (ref 9.1) first refers to SWAN & RAVEN in October 1865 while in the summer of 1866 Henry Brunel describes tests with a pair of five foot model yachts which were rigged and raced against each other (ref 9.2). It is very likely that these were the medium sized SWAN & RAVEN models which, though described as '6 foot' from their overall length were only just over five feet long at the waterline, the significant dimension for yacht designers. By August 1866 Henry is referring to graphs of the resistance of SWAN & RAVEN (ref 9.3).

It is interesting to recall Froude's footnote in his comments on Scott Russell's 1863 paper, briefly mentioned in Chapter 4, in which Froude set out, very fully, the laws by model results can be scaled to their full size values. In this note (ref 9.4) he shows that in resistance to rolling the moment of force will vary as the fourth power of the scale but the moment of resistance will also vary as the fourth power since it will depend on the square of the velocity times the area on which it acts and the lever with which it acts on the ship. He continues

> '..the square of the velocity also (varies) as the scale, since the periodic time is as the square root of the scale.'

Froude then notes that Beaufoy's work (ref 9.5) showed that frictional resistance varied by a power somewhat less than the square and hence that the resistance of the ship to rolling would be slightly less than that of the model.

This note contains every important point used later in his work on resistance to forward motion – velocity scaled as square root of length (Law of Comparison), then residuary resistance is proportional to displacement and finally that friction has to be treated separately because it does not depend on the square of the velocity. It is also clear that the SWAN & RAVEN experiments confirmed a hypothesis which Froude already held rather than leading him in a new direction.

9.2 The Experiments With Swan and Raven

In presenting his results later (ref 9.6), Froude stated his objectives but, as described above, his experimental plan evolved from a much more limited investigation and it may be assumed that the objectives also only reached their final form in the course of the work. It is, of course, quite common for the objectives and the research itself to evolve together as chicken and egg. As stated the objectives were: (ref 9.7)

(a) To compare the resistance, over a range of speeds, of two models of very different shape. One was based on Scott Russell's wave line approach, with trochoidal waterlines and was referred to by Froude as the 'Sharp' model or RAVEN. The other had much bluffer ends, copied 'by eye, not by any attempt at measurement from those of water birds when swimming' and was referred to as the 'Blunt' model or SWAN.

(b) To compare the resistance of models of each shape but of different dimensions, pairs of SWAN and RAVEN models were built with nominal lengths of 3, 6 & 12 feet, each pair twice the length of the next smaller, Froude sometimes referring to the models as Young, Tin & Old. The 6 ft model was made of tin plate, the others were wood, lending support to the suggestion that the six foot models were made at a different time. The majority of runs were run at constant displacement but a few were run at deeper draughts.

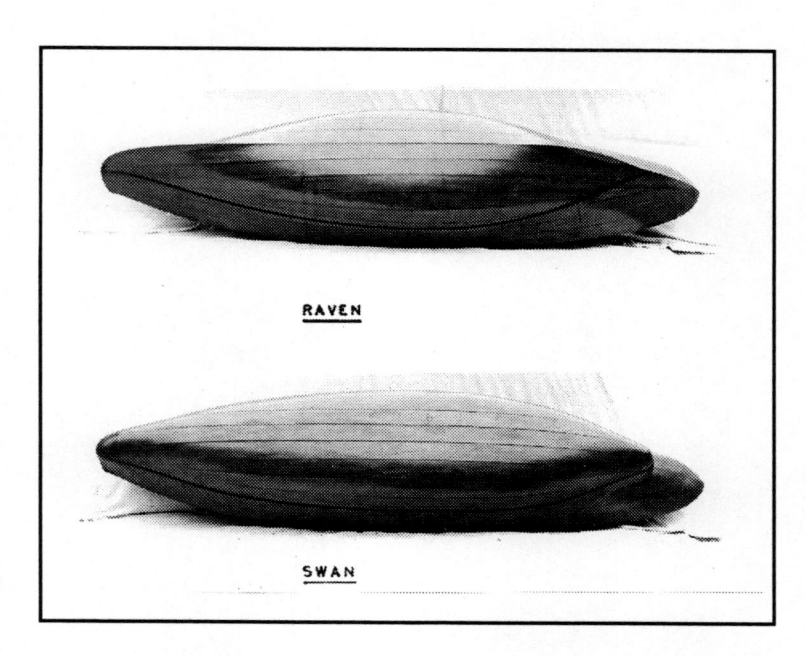

Figure 9.1 – The 'Three foot' SWAN and RAVEN models used to establish Froude's Law of Comparison.

9.3　Experimental Method

The models were towed from outriggers 10ft 6ins from either side of a launch lent by G. P. Bidder, a fellow member of the Royal Dart Yacht Club, an eminent civil engineer and a member of the 'Existing Knowledge Committee' (ref 9.8). The measuring apparatus (fig 9.2), shown below, allowed the resistance, measured against a spring balance, to be recorded continuously

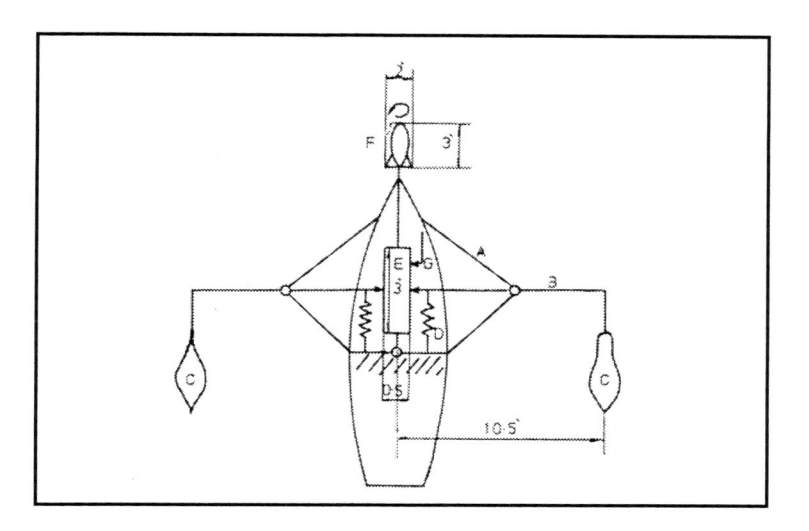

Figure 9.2 – SWAN and RAVEN towing recording apparatus.

on a revolving drum.

Speed was measured on what Froude refers to as a 'Massey' log, though it seems to have differed greatly from the original design. Froude's log was three feet long and two feet in diameter and was carried in a frame just ahead of the launch. Like all Froude's apparatus, it was beautifully made and the rotor, when out of the water, would turn in a moderate breeze. When in place, it would sometimes measure a movement of the launch quite imperceptible to those on board. This log was calibrated by running the launch over a measured distance of 400 feet with 'many intermediate stations at known intervals.'

The rotation of the log turned the drum on which the resistance was recorded by pens, one for each model. A timing apparatus moved another pen along the axis of the drum so that the slope of the line recorded gave a measurement of speed. There are indications of early work on this apparatus from 1863 and it was probably in near final form by 1865 and was used for calibration and preliminary tests up to 1867.

Froude seems to have anticipated considerable difficulty in keeping the launch on a straight course in the current. If the launch was on a curved path the outer model would, of course, move faster than that on the inner boom but he convinced himself that this problem could be overcome. A more difficult problem was that the current and tide were variable and might affect the model on one side more than that on the other. To overcome this problem, the models were interchanged from time to time and the experiments were either run in clear water in Dartmouth Harbour, where the current was uniform, or in a creek, probably Waterhead, where at slack water, there was no current. In June 1866, he used Churston Cove, Torbay, for tests, possibly of seakeeping, and associated with his work on yachts.

Particulars of the Models – Table 1		
(a) True dimensions of the 6ft models (Wright)		
	SWAN	**RAVEN**
Length overall, ft	5.767	5.776
Length, waterline	5.333	5.683
Beam, max	1.333	1.317
Draught	.653	.722
Depth	.871	.854
Volume of displacement, cu ft.	1.287	1.283
Surface area, sq ft.	4.640	4.675
(b) New 12 foot models were re-tested in the tank in 1877 and later the results were analysed and presented by R. E. Froude in his 'Circular' notation. (Chapter 14)		
(M) Length	5.937	5.955
(B) Beam	.947	.954
(D) Draught	.355	.357
(S) Wetted surface area	6.480	6.480
Prismatic coefficient	.553	.548

9.4   Results

The original diagrams are missing, probably destroyed during World War II in the bombing of Devonport Dockyard where many Admiralty Experiment Works' drawings had been sent for safe keeping. The best known plotting is that given to the British Association. This shows that for all sizes the SWAN had the lower resistance at high speeds and was generally more resistful at low speed. (There is an anomaly in the 3ft results, discussed a little later). Froude's first objective was clearly satisfied without reservation; the belief in a universal form of least resistance which had been accepted since Isaac Newton's days was incorrect and shape would have to be selected with regard to operating speed. Writing his "Explanations", he made a further point.

'....these experiments show that strange forms may possess merits that are entirely unknown and unexpected before an experiment is made upon them; for here we find that an abnormal form (suggested simply by the appearance of water birds when swimming), if moving with a high though not excessive velocity, experiences considerably less resistance than the waveline form, the accredited representative of the form of least resistance, particularly at high speeds. This proves that we have no ground for certainty that we have found even an approximation to the best form, unless we have gone experimentally over almost the whole ground and tested a very wide variety of shape. ....on very many important questions, such as, for instance, the proper ratio of length to breadth, there is no really established principle of judgement on which reliance can be placed. Yet most weighty considerations affecting economy and efficiency are involved in the settlement of that single question. But unless we build more experimental ship-sized models, there seems no possibility of determining the question by full-scale experiments.'

This result alone was surprising to most contemporary ship designers. It was reasonable to consider full scale tests on a single 'optimum form' but, once it was accepted that the best form for each ship should be different, the cost of full scale trials would clearly be prohibitive.

Froude's second objective was also achieved, though perhaps less conclusively, as the speeds at which the resistance curves of SWAN and RAVEN cross are almost proportional to the square roots of the lengths of the models and the general characteristics of the curves are very similar. It is clear that most contemporaries came to accept these tests as verifying 'Froude's Law' which he stated as:

'.....for similar models the resistance diagrams should be similar at velocities directly proportional to the square root of the dimensions and that at such velocities the resistance should be as the cube of the dimensions.'

Note that he is referring here to total resistance and does not allow for the fact that friction increases at a power of speed rather less than two; a point which he had made in earlier in responding to Scott Russell's 1863 INA paper.

In the conclusions of his 'General Aspect of Results' (ref 9.9), Froude again shows his powers of observation. He describes the action of the blunt bow as follows.

At these velocities when

'the resistance of the Blunt model is in excess, is that under which there is formed at her bow, a small humpy wave which has no counterpart on the bow of the Sharp model. At the period when the superiority of the Blunt model begins to assert itself, as shown by the round curvature of the resistance line, this little humpy wave begins to lose its individuality and is merged in the full length wave of midship water depression & water elevation forward and aft which is just then assuming a definite shape.

.... The wave she makes is that of a longer model.'

which is remarkably similar to modern descriptions of the action of a bulbous bow at high speed, even to the penalty at lower speeds. His son, Edmund, made a similar point in 1886.

> 'The principle of the SWAN bow may be said to consist in the use, instead of an ordinary form of specified length, of an ordinary form of greater length brought back to the specified length by 'snubbing' at the forward end. The form thus produced possesses, so far as the transverse wave making resistance is concerned, nearly the full advantage of the extra length of the 'unsnubbed' form – at the expense of the extra resistance due to the surge and diverging waves, made by the extra blunt 'snubbed' entrance, and at the speeds at which the transverse wave-making becomes excessive the gain on the former account greatly outweighs the loss on the latter.'

William Froude was also greatly concerned over the anomalous behaviour of the 3 foot models whose curves differed somewhat in character from the larger models. He blamed that of RAVEN, in part at least, on a deadwood fitted for earlier tests and, apparently, at least some of the tests with these small models were made from a hand propelled boat, whose speed would be more difficult to control than that of the launch. The general excess of resistance of these models was ascribed to the 'Rigidity of water'. He also noted the different appearance of the bow waves of the smallest models compared with that of the larger ones. This difference is easily seen today in naval war films where the use of models is immediately obvious from the different characteristics of the bow wave and the size of the spray droplets generated, largely due to the influence of surface tension.

9.5 Hindsight

Wright's replotting of the original curves shows the discrepancies more clearly and it is likely that they were bigger than Froude realised. Since the original experimental values are lost, it is not possible to see what scatter there was before Froude drew the mean lines. A modern experimenter would not be surprised by these difficulties; even a 12 foot model is seen as on the small side and reliable quantitative results would not be expected from 3ft or 6ft models, particularly at the lower speeds. Such small models are used for qualitative work and in this sense, Froude's work is clearly valid. The reasons are highly complex but, at some risk of over simplification, may be summarised as follows.

The effects of viscosity on resistance depend largely on the Reynold's number, which is defined as the product of length and speed divided by the value of the kinematic viscosity. At small

Chapter 9: SWAN and RAVEN Experiments, 1865–1867

size and low speed (Low Reynold's number) the flow in the boundary layer close to the hull, where the effects of viscosity are concentrated, will be much smoother (laminar) than at higher values when the flow is turbulent. There will be a patch of laminar flow on all these models close to the bow but its extent will be greater relatively on the smallest models reducing their resistance, contrary to the overall effect measured. It is likely, though not certain, that laminar flow will be more extensive on SWAN than on RAVEN due to the pressure gradients round her bluff bow which might account for her 3ft model having less resistance at the lowest speed.

Near the stern, the flow will separate from the model, increasing the resistance considerably. Such separation will occur further forward in the smallest models which seems the most likely explanation for the generally high resistance of the 3 ft models. It is not easy by inspection to say which shape would suffer most. Separation is triggered by excessive convex curvature and there is no doubt that the bluff stern of RAVEN would do this. On the other hand, there is a sharp curvature further forwards in the lines of SWAN which might initiate earlier separation. There are indications in the excess resistance of the small SWAN model that this was indeed so. Froude was clearly dissatisfied with the results for this model and ran 53 experiments as opposed to 13 for the RAVEN model. He showed the maximum and minimum values as well as the mean which is another indication of irregular flow such as might be caused by separation. Such uncertainty is one of the reasons why small models are not used today.

An even bigger problem was revealed in 1877 when new 12 foot models were tested in the experiment tank at Torquay. Their resistance was very much lower than that measured in the earlier tests and it appeared that the total resistance associated with viscosity appeared to be almost double that due to friction, an improbable result for forms of this character. A possible explanation, though there is no direct evidence, is suggested by Froude's claims for the test procedure in the tank at Torquay. The models were attached to the carriage which ran on rails so preventing any yawing as they moved along the tank. Models pulled from a single point, as were SWAN and RAVEN in Dartmouth Harbour, are prone to yaw and, as a rough guide, one degree of yaw adds 4% to resistance of a modern form. Shedding of eddies as a result of flow separation would be very likely to provoke yawing. (ref 9.10)

There were other causes of error which were not recognised by Froude or his contemporaries. He showed that resistance was critically dependent on length but the models at the draughts tested were not strictly in the ratio 3:6:12. There must also be some doubt over the accuracy of his speed measurement. The log was only just ahead of the stem of his launch and may have been affected by the bow wave. The length of his calibration run is also rather short. At 170 ft/min the measured distance would be covered in about two minutes but at higher speeds the time for calibration would be less than a minute.

Froude clearly realised that these tests were an incomplete verification of his procedures and it is to the credit of his powers of observation that he could draw valid conclusions from this evidence. The large percentage errors which occurred amount to a few ounces in measurement taken from a vibrating steam launch in a busy harbour. Froude was to obtain more reliable data confirming his work later during the Ramus investigation (Chapter 13) and his Law of Comparison has been fully confirmed by work over the last century. History is right in seeing these experiments and the conclusions drawn from them as Froude's greatest and most novel work.

9.6 Mr. Froude's Explanations

Though Froude was a member of the 'Existing Knowledge Committee' (Chapter 8) he was only able to sign the report 'subject to the following reservations' He was able to agree with Merrifield's detailed survey of existing knowledge but disagreed strongly over the way ahead. His 'explanations' begins with a layman's explanation of the nature of resistance most of which would be seen as valid today. In summarising this section, Froude's words will be used as far as is possible.

> 'The subject of a ship's resistance is one which I have for many years been independently investigating, both theoretically and experimentally; and I have thus been led to conclusions which are in very material respects at variance with those which Mr Merrifield has placed on record for the committee as representing the existing state of knowledge respecting it, and specially at variance with the consequent recommendations which he has drawn up...' He continued 'Until the draft report was in my hands, I was unaware that "Resistance" was regarded as included in the list of subjects submitted to the committee;...'

He then said that the results which he had obtained were in many aspects incomplete and he had therefore been reluctant to publish them but he saw the main conclusions as fully established and these led to a very different line of action from that in Merrifield's report. One of the main recommendations of this report was for the resistance of a full size ship to be measured by towing it. Froude thought such a trial would be of real value and great interest but:

> 'I shall contend that experiments on the resistances of models of rational size, when rationally dealt with, by no means deserve the mistrust with which they are usually regarded, but, on the contrary, can be relied on as truly representing the resistances of

Chapter 9: SWAN and RAVEN Experiments, 1865–1867

ships of which they are the models; and in order properly to open up the question, so great a variety of forms ought to be tried that it would be impossible, alike on the score of time and expenditure, to perform the experiments with full scale ships.'

Froude then says that it was generally believed that the resistance of a given body will increase as the square of its speed and that, in comparing bodies of identical shape but different size, the resistance will vary as the square of the dimensions since this ratio expresses both the proportion of their midship sections and of their surface area, exposed to friction. If these propositions were true, a single test would give a complete curve of resistance for all speeds for both the model and for ships of the same shape. Since it became known that the actual resistance of a surface ship deviates very considerably from this square law; it has been too hastily concluded that model tests cannot be of value.

He then pointed out that this 'square' law held quite well for submerged bodies, suggesting that the apparent deviations in some of Beaufoy's experiments were due to their very angular shape and to the way in which they were towed. The way in which friction varies with speed also is close to the square of the velocity and he instanced his Torquay water main tests in support of this. It is interesting that, in 1863, he followed Beaufoy in suggesting a power less than two; perhaps the rough surface of the water pipe gave a misleading answer in this respect.

Rankine's work on streamlines is then outlined and Froude suggested that, for smooth shapes, even those with very bluff ends, friction was the only aspect of viscosity which need be considered, at least if the model is more than five or six feet in length. Bodies moving in the surface of the water generate waves and the forces which generate these waves, reacting on the body, produce a resistance which does not obey the simple square laws. Froude said:

> 'I shall show, in fact, that if the velocities of the ship and the model are as the square roots, these excesses of resistance thus arising will be as the cubes of their respective dimensions, a law which, as is easily seen, expresses also the relation founded on those elements of resistance which vary as the square of the velocity and as the squares of their respective dimensions.'

The wording is interesting – 'I shall show' is not the same as a claim to have discovered, and it is likely that he had now become aware of Reech's work. (ref 9.11)

Froude then returns to Rankine's work on streamline flow to show that the wave configuration when controlled by gravity will be similar if the bodies are moving at speeds proportional to the square root of their dimensions. The energy abstracted from the body, which is equal to the resistance, per unit length of wave will vary as the cross section of the wave (mass/unit length) times its elevation, or as dimensions cubed. Froude then turns to the tests which he made to confirm his views, already discussed.

It is surprising that Froude's contemporaries had failed to reach similar results. The principles of scaling results from model to full size had been set out by Isaac Newton in the context of ballistics and further developed by various scientists, mainly French, by the early nineteenth century. Reech had derived the Law of Comparison by 1844 and his work was published by Bourne in England in 1852. Fairbairn's use of models in the design of the Britannia Bridge in 1845 was well known and much admired. Men like Scott Russell and Ditchburn and many others had tried model tests for ships but failed to realise the need to test at 'corresponding speeds' and hence they derived incorrect estimates.

It is sometimes claimed that Froude's work was not original and, in the purest sense, there is some truth in the statement. Froude told the Royal Commission of 1872 that

> 'I have indeed seen extracts from a French work on naval architecture, in which sound views on the subject are expressed, but the opposite view has been almost universally held',

in which he is clearly referring to Reech. Froude's achievement was to show the Law of Comparison could be applied to the results of model tests in order to give reliable estimates of the power needed to drive a ship. This procedure was original, valid and of lasting benefit.

References

9.1   Rough note book, held in the Froudes' Museum, Haslar.
9.2   Brunel Collection, Bristol University, 07159 310566
9.3   Brunel Collection, 07202 070866
9.4   W. Froude. *Remarks on Mr Scott Russell's Paper on Rolling*. London, Trans. INA., 1863. (p.95 collected works.)
9.5   Beaufoy's experiments are discussed at length by Wright, and summarised in: D. K. Brown. *Before the Ironclad*. Conway Maritime Press, London, 1990.
9.6   There are various papers by Froude describing these experiments but the differences are trivial. In this chapter, I will draw on that contained in the manuscript 'General Record' held at Haslar as edited and published by Yoshioka. (9.7) The concluding critiques draws heavily, but not exclusively, on the analysis by Dr Tom Wright in his thesis.
9.7   I. Yoshioka. *W. Froude's Report of Experiments made in 1867*. Bulletin of the Faculty of Engineering, Yokohama National University, March 1975.
9.8   E. F. Clark. *George Parker Bidder, the Calculating Boy*. KSL Publications, Bedford, 1983.
9.9   As 9.7.
9.10 Eddy shedding in the form of a von Karman vortex street depends on the Strouhal number. Forecasting of eddy shedding remains imprecise, to say the least, but estimated values of Strouhal number for the SWAN & RAVEN models suggest that it is a distinct possibility.
9.11 Royal Commission on Scientific Instruction and the Advancement of Science. (The 'Devonshire' Committee) Parliamentary Papers 1874, XXII.

Chapter 10
Admiralty Approval

10.1　Introduction

The story of Froude's negotiations with the Admiralty must be told at some length as the latter is sometimes, perhaps unfairly, criticised for undue delay. Sir Edward Reed, Chief Constructor of the Navy, had been an early and enthusiastic supporter of Froude's ideas on rolling and was invited to dine with Froude by Henry Brunel at the latter's London house on 14 April, 1867. The rolling of the ironclads had given rise to a great deal of ill informed comment which Froude and Reed saw as exaggerated and Froude, at least, thought that the few real problems could be greatly reduced by fitting suitable bilge keels.

At the time of this dinner, the SWAN and RAVEN experiments had barely started and though Froude seems to have outlined his views on model testing, he didn't 'jog the matter'(ref 10.1). Reed visited Chelston Cross on 14th February, 1868 and was given a demonstration of the value of bilge keels using the old model of the GREAT EASTERN and Froude also described progress with the SWAN and RAVEN, just before Froude had to return the launch to Bidder (ref 10.2). Froude wrote later: (ref 10.3)

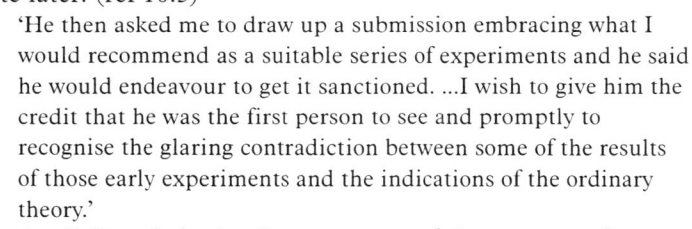

> 'He then asked me to draw up a submission embracing what I would recommend as a suitable series of experiments and he said he would endeavour to get it sanctioned. ...I wish to give him the credit that he was the first person to see and promptly to recognise the glaring contradiction between some of the results of those early experiments and the indications of the ordinary theory.'

By April, Froude had written a report of these tests and a copy was sent to Reed which has been lost together with the covering letter though it seems to have been very similar to the report summarised in the previous chapter.

During the summer of 1868 Henry and Froude had a number of discussions on the way in which the Admiralty could be persuaded to sponsor a model testing tank to extend Froude's work on resistance. During the summer Reed invited Froude, together with his own assistants, Barnaby and Crossland, to dinner at his house in Greenwich at which the proposal was discussed. By 24th June, Henry had written the outline of a proposal which Froude was to follow very closely in a formal submission sent to Reed in December 1868. Henry was concerned that Merrifield's British Association 'Existing Knowledge' Committee, of which Froude was a member, might propose model tests, pre-empting Froude's own bid for support and even taking credit for his work. He suggested (17th August

Figure 10.1 – Sir Edward Reed, Chief Constructor of the Navy, who supported and sponsored Froude's work.

1868) that Froude and Rankine should combine their work on stability and rolling to forestall any action from the British Association. By 24th August, Rankine seems to have agreed with Froude's plans, provided that the tank was made big enough for large models. They also hoped that Reed would be "Commander" of the tank.

During the autumn, Froude had had another meeting with Reed though no details have been found. Froude could hardly have made his proposal at a worse time as both the Board of Admiralty and Reed had other problems on their mind which must have seemed of greater urgency and these will be outlined before considering the details of Froude's proposal.

Edward Reed

Reed's career began in what was then the conventional style for a senior naval architect; first as a shipwright apprentice at Sheerness, winning a very competitive selection for the Central School of Mathematics and Naval Construction at Portsmouth. He was then appointed as a supernumerary draughtsman at Sheerness and, though this was a recognised training grade, he found the routine work boring and resigned in 1853 to become Editor of Mechanics' Magazine. In 1860 he became the first Secretary of the Institution of Naval Architects and submitted a number of his own designs to the Admiralty as a result of which, in 1863, the First Lord, Duke of Somerset, appointed him as Chief Constructor when Watts retired.

Reed produced some outstanding designs, widely copied, and, during the late 1870s, directed or sponsored a number of innovations in the theory of naval architecture. As mentioned in Chapter 5, Barnes and Barnaby of his staff developed the first method for calculating stability at large angles of heel. Reed initiated a design method for the strength of ships, carried out by White, another of his bright young men. He had already adopted Froude's approach to rolling and was willing to support the proposed ship tank.

His experience as a journalist was used to ensure that his ideas were clearly presented, both in technical papers and in more popular works. After his resignation he was joint editor of the professional journal "Naval Science" from 1872–75 and wrote a number of books, both popular and technical, on warship design (ref 10.4).

Developments in Armoured Ships

WARRIOR was designed in 1859 to carry a numerous battery of 68 pounder smooth bore guns, the most powerful afloat. The number of guns multiplied by the length required between them for efficient working, set the length of the battery. Her designer, Isaac Watts, also added the fine ends then seen as necessary, making her a long ship. Most of the battery was protected by 4.5 inch armour which was invulnerable to any projectile available (ref 10.5).

Chapter 10: Admiralty Approval

Table: The development of Armoured Ships

Name	Date Launch	Disp't Tons	Length Ft	Armour ins	No of guns *	Type of Gun	
Warrior	1860	9,210	380	4.5	36	68 pdr	SB
Northumberland	1866	10,700	400	5.5	26	8/9"	MLR
Bellerophon	1865	7,550	300	6.0	10	9"	MLR
Hercules	1868	8,680	324	9.0	10	10"	MLR
Turret ships							
Monarch	1868	8,300	330	7.0	4	12"	MLR
Captain	1869	7,767	320	7.0	4	12"	MLR

* Several ships carried a mixed armament. The number of guns is that of major weapons, excluding saluting pieces. The size of gun given is that of the most powerful piece carried.
SB Smooth bore (Muzzle loading)
MLR Muzzle Loading Rifle

WARRIOR was followed by some ineffective diminutives to save money but her true successors, such as NORTHUMBERLAND, were bigger. More powerful guns were entering service whose projectiles could only be kept out by thicker armour. This led to trends, starting with Reed's BELLEROPHON, for fewer, more powerful guns, permitting a shorter hull with less exposed area needing the heavier armour. HERCULES continued this trend with improved end on fire. The turret ships, MONARCH and CAPTAIN (Chapter 5) may be seen as continuing the trend.

It seemed clear that the trends for fewer guns, thicker armour and shorter hulls would continue and though Reed was prepared to accept some penalty in power for his shorter hulls, he welcomed Froude's ideas as helping to minimise any such penalty. In fact, these ships, which all had top speeds of 14–15 knots, had a speed/length ratio ($\frac{V}{\sqrt{L}}$ Knots/ft) of 0.7 to 0.8 at which at least two-thirds of the resistance would be due to friction. Provided that it was not carried to excess, the shorter ship, with less wetted area, would be less resistant at these fairly low speeds. There are indications that Reed thought that SWAN represented a full form instead of just full ends, though the displacement as well as the length of the two models was virtually identical.

10.2 Childers

Childers was appointed by Gladstone when he took power in 1868 and as a Liberal saw the Naval Lords, Tories except for Robinson, as enemies. He made a number of changes to make his own authority absolute, to reduce the naval element and to cut costs, regardless of the effect on the capability of the navy. Board meetings were almost abolished and business was carried out on paper so that no one other than Childers had the overall picture. Soon after taking office he circulated a memo objecting to the practise of bringing outside political pressure to bear on Admiralty decisions (ref 10.6).

All decisions were his alone and even when such decisions could be justified, they were badly implemented. Childers believed that wisdom was more likely to be found outside the Admiralty and, though reluctant to take advice from his own staff, could be persuaded by glib outsiders.

10.3   Submissions

It was to this divided Admiralty that Froude sent his proposal, with Childers at odds with his whole Board over his way of doing business and, more specifically with Controller and Reed over the CAPTAIN. Froude made a number of interesting points in his covering letter (ref 10.7). He says that the proposal contained the results of the experiments already forwarded, forming the groundwork of a proposal for an enlarged experimental enquiry considered with more complex appliances – 'all this framed in accordance with the suggestions you made when we met in the autumn'.

Froude goes on to say that he would have made the proposal earlier but as it was nearly complete,

> 'there occurred to me a line of thought which appeared to lead up, demonstrating to what is, in fact, the fundamental proposition on which the whole scheme must rest and by which alone it can be justified.'

This line of thought derived from Rankine's basic work on wave theory and on streamlines which enabled Froude to put his empirical results into sounder form. Froude concludes

> '..... if we are really quite sure that experiments on the resistance of a model will tell us truly what will be the resistance of a similar ship..... such an enquiry as I propose is placed beyond a doubt.'

The enclosure was titled 'Observations and suggestions on the subject of determining by experiment the resistance of ships'. It is a long paper and can be read in full in Froude's collected papers (ref 10.8). The paper begins by outlining Froude's views on the difficulties of using full scale trials to develop novel hull forms and on his proof that properly interpreted model experiments can give the information needed to estimate the resistance of ships. It must be repeated that these views, so obvious today, were totally opposed to most 'scientific' opinion of the day.

Froude continued with quite specific proposals for a test tank in which models could be tried. These requirements will be outlined here to show that Froude already had all the essentials

Chapter 10: Admiralty Approval

of a model test tank clear in his mind. There had never been such a tank before and every feature, so clearly set out, was novel. Since the tank was built very much to these proposals, more will appear in the later chapter dealing with the building.

> 'The waterway should be covered to prevent interruptions due to weather and kept at a uniform level, free from scum. The waterway should be deep enough and broad enough so that the resistance was not affected by constriction of the flow and long enough to allow the model to attain its highest steady speed gradually. At this stage he was proposing six foot models which he saw as adequate.
>
> 'A drawing office and workshop would be needed in the same building. Stearine (hard paraffin wax) was already suggested as the material for the models and the design of a model cutting machine outlined.
>
> 'To avoid the problems caused by a model swaying when towed, which had troubled earlier experimenters, he suggested that the model should be attached under the carriage of a light railway, free to move up and down but constrained to move in a straight path. The revolution of the carriage wheels would give a precise measure of distance covered and hence, with a time base, the speed. Resistance would be measured on a dynamometer.
>
> 'Experiments on surface friction would be of high priority after which methodical variations in form and proportion would be investigated.'

Froude offered his services free, pointing out that though there were other men of science who could do the work, their pay would add to the cost. He clearly considered that he was the right man for the job saying;

> 'It is a simple fact, not only that the subject is one to which I have devoted great study for many years, but that I have devoted to it several years of almost exclusive attention, coupled with extensive experiments, carried on at a considerable expense to myself.
>
> 'I have thus acquired a large stock of apposite knowledge and matured habits of experimental inquiry. Indeed I may fairly express a doubt whether any other person has the advantage of me in this respect, or produced as instructive a series of experimental results as that which I have already presented to the Chief Constructor.
>
> 'I am besides, what is termed a good mechanic and I have a good workshop in my own house; I am now providing steam power which will be of material service in the operations I suggest.'

He stipulated that the tank should be close to his house and that he should direct the methodical investigation of resistance and that he should be given funds to pay for the building of the tank, apparatus and to pay the staff. He gave a detailed estimate of the cost of the building at £1,000 and of running the establishment at £500 a year for two years. He was prepared to guarantee these figures, making up any deficit himself. The 'draughtsman and assistant' was to be his son, Edmund, 'who more thoroughly understands the work than any one else whose services I could similarly secure.' He would be paid £3-3-0 per week.

He concluded by reminding Reed that should £2,000 seem a large sum, it could be offset by the reduction of fuel bills from a saving of ten horse power in each ship, or by a reduction in the number of speed trials or by a reduction of ten feet in the length of a single ironclad.

10.4 Consideration

Froude maintained a flow of letters to Reed, perhaps to keep the pressure on him (ref 10.9). On December 27th he wrote asking Reed to agree to him discussing his ideas with the former Tory First Lord, the Duke of Somerset. On January 12th he wrote saying that the Admiralty had returned the wrong models, not the SWAN and RAVEN which Reed had borrowed. In the same letter he said that he could reduce his estimates by £105 allowed for a water main as he now thought he could get the water from a well using his new steam engine to work the pump.

On January 20th he acknowledged Reed's letter of the previous day saying that 'though to some extent discouraged, yet not wholly dismayed by it.' Froude said that his visit to the Duke of Somerset had gone well and if he were First Lord 'there would be little difficulty in obtaining sanction for the proposed system of experiments'. He (the Duke) said that it was a good time for the proposal on the grounds of economy. The Duke's father might present Froude's report to the House of Lords. Reed's letter has not survived but one can assume that it said that money was not available in 1869 as on 25th January Reed wrote to Merrifield:

> 'The present is not a favourable time for obtaining their Lordships authority to increase expenditure on experiments as they are employing every possible and proper means of reducing expenditure.'

Froude wrote again on the 20th:

> 'On thinking over what you say in your letter it occurs to me to ask if it would do good if I were to bring to bear on Mr. Childers any collateral influence simply to the extent of getting myself **statements** guaranteed as a good man and true who has long paid attention to the question. Hitherto it has seemed to me certainly better to leave this aspect of the matter entirely in your hands and it is only as a collateral support to your recommendations that I think of it now.'

Froude continued saying that Mr. (James) Spedding is a friend of Mr. Childers and that some years ago Mr. Spedding assisted me in some model experiments on screw propulsion.

'Mr. Spedding was thus an independent witness of what I was doing and both on his own account and through Lord Houghton could and would make interest for me in the best sense of the word if that would be desirable.'

'The more I think of the matter, the more sure I feel that the sort of enquiry I propose is a measure of the truest economy – but I may well doubt whether I have name enough (you see I don't doubt my merit) to be trusted with such an important job by the public. That is why a collateral recommendation might be of service."

On March 7th Froude tried again:

'A visit from Mr. Spedding added the last straw to the weight which breaks down my hesitation to write to you. Hesitation due to the thought "What an awful bore each additional letter must be" to one whose time is so fully fitted up as yours.

'The Ides of March are come – but not gone. The Estimates are made up but I don't know that my tank and model experiments are included.

'The only thing I have to say, that needs saying, perhaps is that if Mr. Crossland's report is sent in and there are any parts in it which are unfavourable to my proposal, I should be glad an opportunity of discussing it.

'Mr. S. has, of course, said nothing to Mr. C. but is ready to do so whenever the fit time comes.'

Crossland's report to Reed has not been preserved but it would normally have been subsumed into Reed's own report to the Controller. This is a careful and supportive critique of Froude's proposal. As it repeats much of the original, only the passages commenting on the scheme need be reproduced (ref 10.10).

'Froude points out certain possible sources of error in the results of his experiments but he concludes, as I think he is entitled to conclude, that taking into account the numerous experiments made, the difference between the two models may be taken as correct.........

'None of these objections (to full scale trials) lie against experiments made on a small scale. They can be multiplied to any extent without serious cost. They can be made to show the resistance of various types without interference from other elements of inquiry such as the efficiency of propellers or the effects of tide, wind or rough surface. They can be made under cover and can therefore be pursued in all weathers.'

After outlining Froude's description of the tank and apparatus Reed continues:

'Mr. Froude argues that he is from many circumstances specially fitted to prosecute such a series of experiments. With this I entirely agree – he is a gentleman of many Scientific attainments, He has given special attention to questions concerned with Naval

Architecture and has communicated valuable papers to the Institution of Naval Architects.

'His paper on the rolling of ships shed quite a new and original light on that subject and is perhaps the most remarkable paper published in the history of the Institution.'

After confirming the validity of Froude's work on resistance based on SWAN & RAVEN, Reed continued:

'I am of the opinion that he is the fittest person from taste and attainments to make such experiments if the investigation were decided on... I consider it my duty to say that, large as the proposed expenditure is, it would, in all probability, be much more than compensated by the results which would flow from it...'

Up to this point, Reed's comments could hardly have been more favourable to Froude's proposal but in his last paragraph there is an apparent shift. A possible explanation will be given later but the shift may also mark the transition from Crossland's draft.

'but looking at the smallness of the total vote for experiments this year, and to the objections which would probably be raised to the expenditure of the public in the proposed manner and looking also to the manner in which even the momentous improvements in the forms and proportions of Iron-Clads which have lately so successfully carried out under heavy personal responsibility, have been dealt with by the Government, I see no alternative but to submit that Mr. Froude be thanked by their Lordships for having placed the results of his interesting and important experiments so freely before their Lordships' officers, and be informed, with regret, that their Lordships find themselves unable to sanction the necessary outlay on the course of experiments which he has proposed, and which have been laid before their Lordships by the Chief Constructor.'

Froude's proposal would come under the heading of experiments on which the total annual expenditure was usually only a few thousand (Vote 10, Section 10, sub-head E – £3,000 in 1870–71) and hence £2,000 seemed a large sum. A similar approach remains a fallacy today. "Research" is seen as an entity leading to futile arguments over the relative value of sea keeping and electronic warfare. The true debate is between spending on warship construction and on the research which is essential for their design. Such a debate will still be difficult when funds are scarce but, at least, it can be discussed in meaningful terms.

Spencer Robinson took a braver line, sending a minute in early April 1869 to both Childers and the First (Political) Secretary, W. E. Baxter, saying that the proposed experiments 'were of great interest and should be tried.' He recommended that £1,000 be made available immediately, the rest the following year. Baxter then wrote to Childers: (7th April ref 10.11)

Chapter 10: Admiralty Approval

'I have talked over the matter with Sir S. Robinson and having some doubt as to the wisdom of his opinion as noted, some independent scientific authority ..... the Royal Society, might be consulted ..... regarding importance.'

With this minute further discussion within the Admiralty ceased for the time being, partly because it was known that the British Association 'Existing Knowledge' Committee was due to report. This report was received on 29th September 1869 with its recommendation for full scale towing trials and with Froude's strongly dissenting 'Explanations'. It is worth remembering that this was the first public statement of the SWAN & RAVEN experiments.

In the meantime, James Spedding had written to Childers (3rd April 1869). This letter is virtually a character reference and will be considered in Chapter 15. It does not seem to have had a direct effect on the Admiralty and was probably unnecessary as Reed had prepared the way for a favourable decision.

Froude and Henry Brunel had been in frequent correspondence, seeking ways to influence the Admiralty to decide, quickly in favour of the tank. Many of Henry's views are summarised in ref 10.12. He was concerned that, formally, Froude had not approached the Admiralty but had written a private letter to Reed and in a style which did not even demand a reply. He should recapitulate what was in his papers of April and December and would be glad to explain his views to scientists such as Rankine or Wooley, both eminent hydrodynamicists. Froude should make it clear that he was not advocating any particular form of ship but a careful series of experiments which would result in a great saving of money. Henry suggested that Froude should write such a letter and show it to Reed, asking if it would help.

There was a risk in a formal approach as, almost like a Parliamentary Question it would demand a reply from Childers and might 'put his back up' and might even 'make them say no'. He agreed that it would be wrong for Spedding to write again and reminded Froude of Childers memo saying that he would have no political interference. Wooley had apparently told Froude that the hitch was not due to lukewarmness on the part of Reed and his department and the proposal was not shelved or rejected.

Froude did not agree with this line and Henry, in his letter of August 3rd, warned again of possible interference from the British Association. The following day, Henry wrote again, referring to an approach to Childers by Thomas Acland, MP for North Devon, and an old friend of Froude's. Unfortunately, Acland's letter cannot be traced and neither can Childers' reply which seems to have been encouraging.

During August, Henry wrote several times urging that Froude

should resign from the British Association Committee and disassociate himself completely from Merrifield. In particular, his "Observations" on the British Association report should be factual, describing what he had done but not disclosing his ideas for future work. Also in August, Wooley visited Chelston Cross and studied Froude's test results.

Merrifield had written to the Admiralty seeking trials data in July. In September Reed received the British Association report (ref 10.13). The demand for a number of full scale trials was not well received and it is probable that Merrifield was never the threat that Henry supposed.

Round about the end of September, Reed wrote to Froude who passed the letter to Henry. The original letter has not survived but from Henry's reply, (5th October 1869), quoted below, it may be deduced that Reed had said that money had been inserted for Froude's experiments in the draft estimates for 1870.

> 'Observe that Reed speaks of the Estimates so I was right in supposing that the experiments will be included in next year's estimates which I believe means you can't begin work until April. I think and hope all will go well.' (ref 10.14)

In the same letter, Henry refers to a letter from Merrifield to Froude, Henry did not trust or like Merrifield, 'a little minded cuss', and warned, again, that Merrifield might try to take over control of Froude's experiment programme and the credit for it. Froude's views on Merrifield are not recorded nor is there any evidence to support Henry's idea of a plot. Henry also passed some comments on a discrepancy between Acland's information and that conveyed by Wooley over the way in which Froude's work should be directed.

There were still a few formalities necessary before final approval was obtained. In the autumn, Froude wrote to Childers, referring to his reply to Acland only to be told, incorrectly, that his original papers had been mislaid. He wrote again on December 11th saying that Mr. Reed knew where the original proposal was to be found and went on to summarise the main points then saying that

> 'Since the proposal was made I have seen no reason to modify either its details or the views on which it was based though I have given continued reflection to the subject;...'

Froude pointed out that if approval was not given soon, he would have to get rid of his trained staff as he was paying them out of his own pocket.

Childers obviously asked for information on which to base a reply and on 21st December Reed wrote to Spencer Robinson:

> 'The experiments which Mr. Froude proposes to undertake would be of real value to Science and would probably develop results of service to ship construction.
>
> 'The sum to be provided for experiments in the Estimates for

Chapter 10: Admiralty Approval

next year, namely £2000, has been fixed to some degree in view of these experiments and I consider that their Lordships would not only confer a benefit upon Science but probably provide improvements of considerable economical importance and the construction of Warships if they gave their consent to the expenditure of £1000 during the ensuing financial year upon the scheme of expenditure laid down by Froude. (The original is not easy to read) – and there can be little doubt that the committee (BA) itself, notwithstanding the divergence of opinion in this upon the question of models – will nevertheless be grateful to learn that their Lordships have taken the step which I here recommend.'

The Admiralty approached Dr. Wooley, asking his views on the proposal and he replied on January 26th, 1870, having visited Froude in August 1869 and seen his results.

'I own to having formerly shared in the opinion that much practical information could not be obtained from experiments on models. But on very carefully reading and weighing Mr. Froude's remarks on the subject I have seen reason to modify my opinion and I think there is very good reason for believing that much valuable information of a nature immediately applicable to the solution of the important problem of combining economy with a minimum available resistance in ships by apportioning to this a ———would result from the series of experiments recommended by Froude.'

Support from a former opponent of model testing was particularly valuable and Wooley had drafted an Admiralty letter to Froude agreeing to his proposal. It was sent on February 9th, 1870. It was in formal style but contained everything that Froude had asked for.

'It is the desire of my Lords that these Experiments should be carried out in the manner proposed by yourself, and in accordance with a detailed scheme which they have directed Dr. Wooley and the Chief Constructor of the Navy (Reed) to draw up in concert with yourself and to submit for the approval of my Lords.'

The rolling experiments suggested by Canon Mosely were to be included without extra cost, the sums of money estimated by Froude were not to be exceeded and no expense was to be incurred until Parliamentary approval had been obtained.

Henry was delighted and wrote to Froude:

'I larf, I do larf. When I read their Lordships letter I screamed. We couldn't have written it better ourselves if we had spent hours over it. Their Lordships certainly have a keen sense of humour.'

In a sense, Froude and Henry had written this letter; it is normal Civil Service practise in approving a proposal to follow the wording of the original to make it clear what is being approved.

One can only speculate as to Reed's personal involvement in

winning approval for the tank. When the new tank was opened at the National Physical Laboratory in 1910 – to be known as the Froude Tank – he told the INA that

> 'When my attention was first drawn to the tank of the late Mr. Froude, and I proposed the Admiralty should assist him in his work, there was great opposition on the part of the Board of Admiralty; but I pointed out to them it was from the point of view of the shipowner.......that such a thing as this was necessary. I took the liberty of suggesting that it seemed to me very strange that they should have engaged me to design Her Majesty's ships.... and then for me to have to tell them that I had to do it in great darkness and in much ignorance, and with many shortcomings, because there was no organised knowledge upon which we could proceed in those days.' (ref 10.15)

This is reasonably consistent with Edmund Froude's account. (ref 10.16)

> 'In reference to the initiation of the Admiralty Experiment Works, he thought it ought to be said that, next to his father, this was owing to Sir Edward Reed, who had been the Chief Constructor at the time when his father had been making the original experiments at his own cost. Sir Edward Reed had been greatly struck with the results obtained from these early extemporized experiments with ship models; and had suggested to his father, who would not have been bold enough to think of such a thing himself, that he should draw up a proposal and estimate for an experimental establishment. That proposal had been carried through, and had been the origin of the Torquay establishment.'

Reed's statement to Controller in March 1869 can then be taken at face value: there was no money in the 1869 Estimates and approval could not be given. He may even have hoped that he would be over-ridden and money found. However, he seems to have ensured that there was money inserted in the Estimates for the following year as he told Henry in October. Reed was in a very difficult position, savings were to be made regardless of the long term costs and he probably felt that a cautious approach would be most likely to achieve results.

Froude's first formal proposal was made in December 1868 and was finally approved only a little more than a year later, in February 1870. For a project rejected by most of the scientific establishment, at a time when the senior men of the Admiralty were at odds with each other, this does not seem unduly dilatory. An equivalent proposal today would be unlikely to win approval so quickly.

The Admiralty's approval of Froude's model test programme was not greeted with much enthusiasm; indeed, when Merrifield spoke to the INA in 1870 there was almost universal condemnation. It must be remembered that the British Association report with Froude's "Explanations" describing the

Chapter 10: Admiralty Approval

SWAN and RAVEN tests had only recently been published and it is likely that few of the objectors had read it. (ref 10.17)

Merrifield opened by outlining the main report of his Committee report listing the members as C. W. Merrifield, G. P. Bidder, Capt Douglas Galton FRS, Mr. F. Galton, Prof Rankine, and W. Froude, and said (Correctly) that the majority accepted 'my view of giving preference to experiments on full sized models.' He had written to the Admiralty on 27th September 1869 and had received a reply dated 9th February 1870.

> 'With reference to your letter of the 28th September, 1869, I am commanded by my Lords Commissioners of the Admiralty to inform you that, after full consideration, they are unable to give a general assent to the proposal of your Committee to conduct experiments on Her Majesty's ships in the fiords of Norway or on the inland waters of the West Coast of Scotland; but my Lords have been pleased to sanction certain experiments on models, to be conducted by Mr. Froude, a Member of the Committee, and will cause the results of those experiments to be communicated, when complete, to the Institution of Naval Architects, the British Association, and such other professional bodies as to my Lords may seem desirable.
>
> I am, Sir,
> Your obedient Servant,
> Vernon Lushington'

Merrifield then said that:

> 'I feel that I can acquiesce with very good grace in the substitution for them of a set of valuable detailed experiments on models, conducted under such superintendence as we may expect from Mr. Froude.'

It must be unclear how Merrifield could say this since he had listened to Froude's views on the Committee and rejected them. The Chairman was Scott Russell and far from impartial:

> 'We are all concerned that such experiments [full scale] were wanted, and that the Admiralty was the right party to make them.'

He said that he was disappointed though pleased that the Admiralty had shown goodwill to the science of Naval Architecture –

> 'I do not announce to you that experiments on very little models are very safe data for experiments on large ships...... I have taken trouble to make a series of experiments on 120 small models extending from 24ins to 12ft and from 30ft to 60ft.'

The only result was to show that large size was needed.

> 'Therefore you will have on the small scale a series of beautiful, interesting little experiments, which I am sure will afford Mr. Froude infinite pleasure in the making of them as they did to me, and will afford you infinite pleasure in the hearing of them; but which are quite remote from any practical results upon the largest scale.'

Scott Russell said that he had made a model 100 feet long of Newton's body of least resistance but when tried it had the highest resistance. This was hardly surprising as Newton's approach was seriously in error.

There was a confused debate in which it was clear that few knew what Froude had proposed. Reading between the lines, it would seem that Froude had not expected an attack at this meeting and had not prepared a defence. Edwin Henwood and C. F. Henwood got muddled with shallow water effect and rough weather but referred with respect to Robert Rawson's experiments and, of course, to Newton.

Charles Lamport provoked Froude into debate by saying that he 'greatly regretted that the Admiralty are going to throw away money........small experiments are useless' and asked what size Froude intended. He was muddled over the relationship between the size of the model and that of the tank fearing that the interaction and "wave of translation", problems which affected canals might apply to Froude's models. Froude replied:

> 'The proportions which I propose would be similar to that of a frigate, with a channel about 2000 feet wide and 33 fathoms deep. That will be about the proportion of the largest of the models used and the canal used.'

G. B. Gallow merely stated that models cannot be relied on. Froude then made a lengthy reply, much of which can be omitted as having already been presented in this chapter. The important points were:

> 'I see that the feeling of the meeting is very much against experiments with models, but I must say that my own experience leads me to judge quite differently.
>
> 'Attention has not been paid to the relation which should subsist between the speed at which the model is moved and the speed at which the ship is moved. No doubt that if you draw the model at some random velocity and ascertain its resistance and conclude that you have measured the resistance of the ship for all velocities you will be quite wrong.'

Froude drew attention to the need for careful apparatus, extra patience and extreme persistence and said that he could not accept the broad conclusions of Scott Russell. Even in print, one gets the impression that, for once, Froude was on the point of losing his temper.

As another Barnaby was to say in the Centenary History of the INA

> 'Unanswerable as Froude's statements seem today, (1960) they did not appear at all convincing to the audience of 1870, and these "minor experiments," as Scott Russell termed them, were thought a poor and misleading substitute for full-scale trials with actual vessels.' (ref 10.18)

No wonder the Admiralty took some time over its approval.

Chapter 10: Admiralty Approval

Referemces

10.1 Henry Brunel correspondence. 12th February 1868, No 398

10.2 Henry Brunel, 22nd February 1868 - also 27282 080679. Note Brunel letters unless very important will not be individually referenced if it is clear from the context and date where they come from.

10.3 W. Froude. Reply to discussion on Experiments on the effect produced on the wave-making resistance of ships by length of parallel middle body. Trans. INA., London, 1877.

10.4 D. K. Brown. The Era of Uncertainty. in Steam, Steel, Shellfire. Conway Maritime Press, London, 1993.

10.5 D. K. Brown. Limits to Growth, British Battleship Design 1840–1904. Conference 'Five hundred years of nautical science'. National Maritime Museum, Greenwich, 1979.

10.6 N. A. M. Rodger. The Admiralty. Terence Dalton, Lavenham, 1979. Note that T. Ropp, Development of a Modern Navy, Naval Institute Press, Annapolis, 1987, gives a somewhat different view of Childers.

10.7 Public Record Office, 116/137.

10.8 The Papers of William Froude. INA., London, 1955.

10.9 As 10.7.

10.10 As 10.7.

10.11 As 10.7.

10.12 As 10.1, No 351.

10.13 Report of a Committee on the state of existing knowledge on the Stability, Propulsion and Sea-going Qualities of ships. British Association. 1869.

10.14 As 10.1, 5.10.69.

10.15 Sir Westcott S. Abell. William Froude. Devonshire association, 1933.

10.16 R. E. Froude. Experimental apparatus at AEW Haslar. Trans. I. Mech. E., London, 1893.

10.17 C. W. Merrifield. The experiments recently proposed on the resistance of ships. Trans. INA., London, 1869.

10.18 K. C. Barnaby. The Institution of Naval Architects. INA., London, 1960.

The Way of a Ship in the Midst of the Sea

Chapter 11
Mr. Froude's Tank at Torquay

1870–79

11.1 The First Ship Tank

This was the world's very first ship model test tank; every feature of the tank and its equipment was novel as were the operating procedures. Froude's proposal of 1868 had outlined his plans for the tank and these were followed closely during its construction though small changes were made to overcome minor problems which became apparent as the building progressed.

Earlier model experiments had been carried out exposed to the weather, in docks, canals or, as in Froude's own earlier work, in estuaries. Working in the open can waste a great deal of time and can also lead to unknown inaccuracies due to the extra resistance from wind and surface waves together with the effect of currents. Rain can affect instrumentation as well as ruining note books. Until 1960 when a large covered tank was built the Admiralty Experiment Works had to conduct manoeuvring tests in an exposed, artificial lake and it was found that useful data could only be obtained on one-third to one-half of working days, to which one must add travelling time if the site is distant – as it was for the SWAN and RAVEN tests in the Dart. There could also be problems with scum forming on the surface of stagnant water.

In earlier experiments the models were towed from a rope and were free to yaw (Chapter 9) and, in most cases, the model was

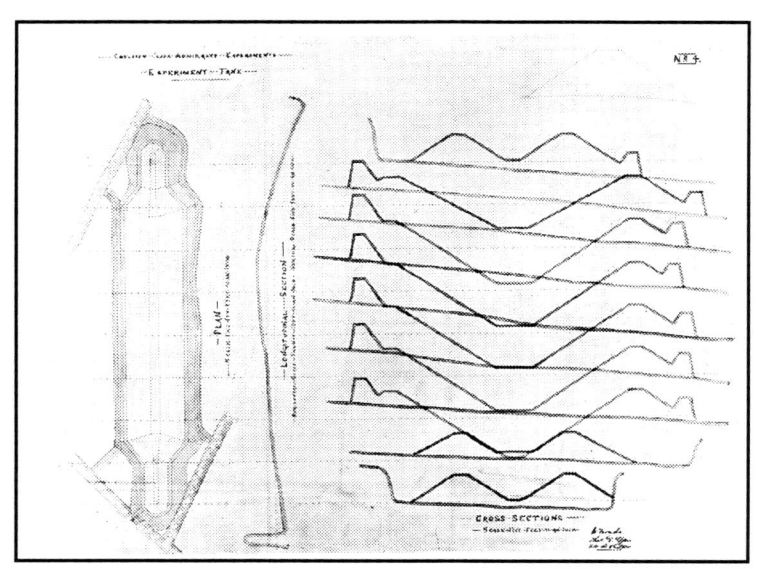

Figure 11.1 – The site plan with sections for the building of the tank at Torquay.

pulled by a falling weight, applying a constant force and the steady speed was measured. Froude's tank was different; the model was suspended under a carriage, prevented from yawing, which can add appreciably to resistance, and driven at a pre-determined steady speed, measuring resistance. Basic resistance tests were, and are, carried out in still water for consistent results. To a first approximation, a form which has low resistance in calm water will also have low resistance in waves. Froude suggested some modifications to hull form for good rough weather performance and, today, further changes are made but the effect is not great.

Froude's apparent simple ideas of covering the tank and constraining the model to run straight at constant speed greatly reduced the time needed for a set of experiments and at the same time increased the accuracy of their results and has been followed in almost all later tanks.

11.2 The Design of a Ship Tank

The overall length of a ship tank is made up of lengths required for:
- acceleration from rest and settling to a steady speed
- enough time and distance at the highest steady speed for sufficient readings to be recorded
- braking to rest without damage

The models must run at the corresponding speed (Chapter 9) which depends on the square root of the length and hence bigger models will need a longer tank. Froude expected to use 6 foot models for most tests, accepting that the occasional 12 foot

Figure 11.2 – The interior of the tank as completed. (The identity of the man on the walkway is unknown.)

model might be limited in speed and hence chose a length of 250 feet for his tank. As built, the Torquay tank was 278 feet long with 195 feet at full width for the steady run, narrowing at both ends, later found to be an unfortunate feature. It was known from experience in canals that the proximity of either sides or bottom could affect the resistance of a moving vessel and hence Froude made the test length 36 feet wide at the surface, tapering to 6 feet at the bottom. The maximum depth of water was 10 feet.

The model was drawn through the water at a maximum speed of 1250 feet/minute by levers from a carriage, or 'truck' as Froude called it, running on rails. It was pulled by an endless cable running over a driving pulley which was worked by a steam engine. The rails had a 40 inch track, 20 inches above the water and were supported off the roof trusses (fig 11.2) to avoid any interference by supports in the water with the waves generated by the model. There were pillars supporting the roof which did stand in the water channel but these were well to the side, 8 feet apart and far from the model.

11.3 The building

The tank was built in a field just the other side of Seaway Lane from Froude's house, on land leased at £12 per annum from W. Mallock, married to Froude's sister and whose son, Arnulph, was one of Froude's team of helpers. The shape of the site led to the tank running South to North, runs starting at the southern end. It was envisaged that the experiment programme could be completed within the two years of Froude's proposal to the Admiralty after which the land would be cleared as required in the lease. In consequence, the main building covering the tank, workshops, offices and drawing office was of a cheap and flimsy wooden construction with a brick boiler house adjacent. It was to last 16 years, with difficulty, but the terms of the lease were fulfilled, the site was cleared and only the memorial plaque on Seaway Lane shows what was achieved there.

11.4 Construction

Froude's proposal of 1868 refers to a tender of £584 for the tank and building and this was updated by a local firm, Bragg and Dyer of Paignton, on 29th April 1870 to £598-10-0. The contract was placed on 21st June and work began at once. Froude covered 30 pages of foolscap in his handwritten specification for the work which clearly drew on his experience as a railway construction engineer and reserved all decision making to himself.

The equipment and apparatus had to be designed and built at the same time as the main building work. A 'time sheet' (ref 11.1) for the late summer of 1870 suggests that Beauchamp Tower was almost entirely occupied with the two steam engines and with day to day correspondence with contractors while Edmund Froude was busy with the tank itself and with the 'office'. Messrs E. V. Acrell, C. Manning and G. Kingdom, probably mechanics, were employed on a range of tasks, primarily the manufacture of the engine but including a well and work on Froude's house.

The first task was to produce 296 detailed guidance drawings developing Froude's concept. It is likely that most of the drawing work was carried out by Beauchamp Tower.

	£
Steam hauling engine (no. 2)	49
Experimental tank, building, and boiler house	40
Truck	39
Screw truck (propeller dynamometer)	32
Model shaping machine	60
Moulding box	5
Mr. Brunel's Section machine (graph paper ruling)	13
Wave making apparatus	10
Miscellaneous	48
TOTAL	296

Many of these drawings are preserved in the Froudes' Museum collection at the Science Museum, Wroughton. (Samples shown in fig 11.3)

It was originally intended to make the tank watertight using 'puddling', a mixture of clay and sand, as used in canals by Brindley and later engineers. Froude sought advice on puddling from Henry Brunel and others and also carried out some tests of his own during June 1870 as a result of which he decided in

Figure 11.3 – The original drawing for the carriage.

Chapter 11: Mr Froude's Tank at Torquay

December to protect the waterline area, which would be washed by waves, with 'asphelte' (sic). By August the excavation was finished and materials for puddling were 'reserved' but puddling could not commence until November as Devon was suffering from an 'unexampled drought' calling for restrictions on the use of water in Torquay.

The tank was filled in January 1871 but was found to be leaking and it was drained to find out why. The problem turned out to be holes made by earthworms which had been expected only near the surface and the puddling there had been salted to discourage them. It was now decided to salt all the puddling and to cover the whole inner surface of the tank with asphalt. This worked as when the tank was refilled in February there was no sign of leakage.

Froude's proposal had included £105 for the cost of water but on 27th December 1868 he had suggested that he could dig a well which would provide water without cost. Though some work was done on the well it was unsuccessful and the tank was joined to the main by a pipe. The water was supplied free in recognition of Froude's services to the Torquay water company.

The main building was of light wooden construction with a felt roof which it was intended should be waterproofed using a patent composition but this was unsuccessful, the roof had to be boarded and covered with tarred felt. The exposed woodwork was also tarred. By October the main structure was complete, the windows were in place and the building was finished by November after some delays due to the heavy rain which followed the earlier drought. The contract was settled in December for £619, slightly less than the original tender increased by £20-10-0 for agreed extras covering the work on the roof, alternative ventilators, a raised and lengthened workshop, different excavation and an increased number of lights.

It was recognised that this tarred building was a fire risk and the boiler house was built of brick by Bragg & Dyer some distance away. The chimney was 30 feet high and a further 80 feet away, on the high side of the suite joined to the boiler house by an underground flue. Insurance was arranged with the North British covering the full value estimated at £1,200 for a premium of £6-6-0. The policy was in Froude's name but, after discussion, the policy was endorsed to the effect that the property and effects belonged to the Admiralty.

There were still problems: Froude's report for January 1871 says 'I may add that some delay has been occasioned by my necessary absences in London during the sittings of the 'Committee on Designs'. The May report elaborated on this problem, '...some important experimental work which has devolved on me in connection with the proceedings of that Committee have unavoidably delayed operations here and the work has been very ably conducted by my son and my assistant,

Mr. Tower, during my absence yet both their time and that of my workmen has been occasionally diverted to the preparation of experimental apparatus for the Committee's work.' Froude was also away during October 1871 for the GREYHOUND trial. Henry Brunel, in a letter dated 9th April 1871 was inclined to blame Froude for the delays (ref 11.2). He thought that Froude was to some extent slow and the blame was to a great extent due to the desultory supervision Froude exercised and he would be able to point to cases of disobedience. Henry thought that Froude did not direct the design work properly and that during his frequent absences it ought to have been the duty of his subordinates to have drawings and designs ready so that when he had time to spare he could approve them. The drawings were often carried in people's head and it was a mercy that things got done at all. Henry went on 'Froude often thinks he is giving an order when his words only imply expressing an opinion'.

Henry had very little to do with the design or building of the tank. He was very busy in 1870–71 and though he kept in touch with the work through frequent letters he can have had little direct experience of Froude's working on this project and his comments must relate, at least in part, to earlier experience. Against this, Froude had, though 30 years earlier, been a tough and capable railway construction manager. Though William Froude was gentle in manner his meaning was very clear and his workmen had been with him many years and knew what he wanted and, in turn, he was satisfied with them. Henry also devised a complicated but apparently practical method of getting letters through quickly. He worked out that a letter posted in London at 4:15 am would catch the 5:30 train arriving at Torquay at 1:44 pm which would reach Froude by 2:30 pm and, if he replied by 7:30 pm, Henry would get it the next morning.

The correspondence book (ref 11.3) lists some 105 suppliers of whom 56 were in South Devon. Correspondence was handled either by Edmund or Beauchamp Tower and there is little sign of replies having to await William's return. One or two items relied on personal favours from William's friends and these may have been delayed while he was away. For example, the rails on which the model cutting machine ran were ordered from the Chief Engineer of the Cornwall Railway, P. J. Margary, William offering another five shillings (25p) if the foreman would pick out a pair of straight ones. These were then machined at the GWR works at Swindon, arranged through another old friend, Sir Daniel Gooch.

Chapter 11: Mr Froude's Tank at Torquay

11.5 Equipment

William Froude, with some help from Edmund and Beauchamp Tower, designed the whole of the equipment for the tank but their drawings showed the major features only and the manufacturing drawings were produced by the contractor. The most important contractor was G. D. Kittoe of Kittoe and Brotherhood (Clerkenwell) who had worked for I. K. Brunel at the same time as Froude (see Chapter 2). Kittoe was to complain (ref 11.4) that the price agreed with Froude for the work was too low for his man to visit Torquay during installation and testing. All the equipment design work had been finished by August 1870 which must have proved a severe load on the small team during the building of the tank itself.

When the Dutch engineer, Tidemann, asked if he could have drawings of the apparatus in 1875 Froude referred him to Kittoe as there were no detail drawings at Torquay. Dirkzwager (ref 11.5) is almost certainly mistaken in suggesting that Froude's reply was deliberately unhelpful on the grounds of secrecy. In his reply Froude said that Kittoe was 'An admirable craftsman and sketcher and a thorough mechanic who would I am sure be ready at no great cost to make a list and sketches and see to the things required.' (23rd Nov. 1875).

11.6 Carriage Rails and Dynamometer

The carriage or 'truck' was a simple wooden box, (figs 11.4, 11.5) strong and shallow, with an axle at each end carrying the large iron wheels. The wheels on one side only were grooved to

Figure 11.4 – The carriage at the beginning of a run.

The Way of a Ship in the Midst of the Sea

Figure 11.5 – Another view of the carriage. Note that only the wheels on the further side are flanged.

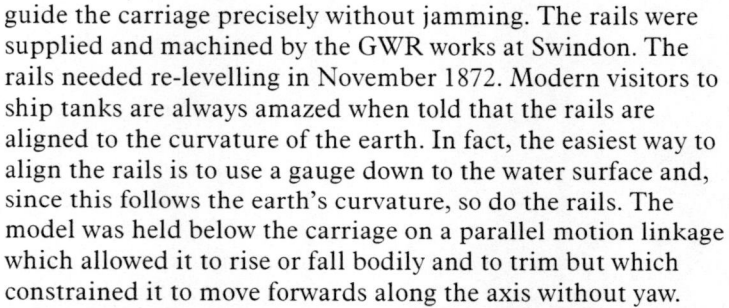

Figure 11.6 – The lever used to transmit force measurements. (Now displayed in the Science Museum.)

guide the carriage precisely without jamming. The rails were supplied and machined by the GWR works at Swindon. The rails needed re-levelling in November 1872. Modern visitors to ship tanks are always amazed when told that the rails are aligned to the curvature of the earth. In fact, the easiest way to align the rails is to use a gauge down to the water surface and, since this follows the earth's curvature, so do the rails. The model was held below the carriage on a parallel motion linkage which allowed it to rise or fall bodily and to trim but which constrained it to move forwards along the axis without yaw.

The dynamometer measured the horizontal force needed to overcome the resistance of the model at the set speed. The extension of a spring was transmitted by levers to a pen which moved over a rotating drum driven off the axle. A separate pen, worked off a clock, recorded a time base on the same drum, with a mark every one and a quarter second as a spring hammer disengaged from a tooth wheel. A large and clumsy looking lever locked the model and carriage against a stop during acceleration while a similar lever locked under braking.

A bell crank lever was fitted, from which weights could be hung to calibrate the spring movement. Springs of different stiffness could be fitted to measure over the range of resistance forces at a convenient scale. This dynamometer was a more difficult job than Froude expected as he wrote in May 1871 'I must add that the difficulties encountered in the reduction of the Dynametric apparatus into working form have been less easy to overcome that had been expected.' There were still problems and in December 1874 the lever system was re-designed. The main parts of the original lever system are preserved in the Science Museum where it may be seen that the

Chapter 11: Mr Froude's Tank at Torquay

levers are beautifully made, light truss girders designed by Froude and made by Kittoe. By July 1872 measurements of trim were also being recorded. Trim could prove a problem to the experimenter as it was found in 1873 that when the bow trimmed down, as in model designated AE (ref 11.6) the forward chucking beam hit the buffer.

The carriage was drawn along the rails by a quarter inch endless wire rope. The original rope had to be sent back (ref 11.7) as it was not in one piece and the full 830 feet was needed in a continuous length.

11.7 Engine and Governor

There were two steam engines, one to haul the carriage and the other to operate the model cutting machine and other workshop equipment. Both were designed by Beauchamp Tower and built in the Chelston Cross workshops. The hauling engine was 'specially constructed to afford a very uniform tractive force.' It was a high pressure, twin cylinder engine with cylinder diameters of 4 inches and stroke also 4 inches, fitted with a heavy flywheel (see fig 11.7). The steam was regulated by a Baily patent reducing valve and then throttled by the governor. In January 1871 it was complete and, together with the hauling drive, made by Sir J. Whitworth, was fixed in place. Tests were

Figure 11.7 – The steam engine (designed by Beauchamp Tower) and winding gear which pulled the carriage.

The Way of a Ship in the Midst of the Sea

carried out successfully in February and the workshop engine was complete at about the same time. The speed range could be selected by using interchangeable gear wheels. Equal speed wheels gave a carriage speed of 180 ft/min at 150 rpm and 330 ft/min at 350 rpm. Other ratios enable the carriage speed to be varied between 40 and 1250 ft/min.

The governor was originally a modification of the common Watt type but Froude improved the design and later further modified it. The Froudes, father and son, were fascinated by governors and three survive in the Museum collection at Wroughton, all much modified (fig 11.8).

Figure 11.8 – The governor which controlled the speed of the engine and carriage. Both William and his son were fascinated by governors and these were frequently altered or replaced.

In the usual governor, centrifugal force will cause the weights to move out and up, closing the steam valve as speed increases. In Froude's design there was an additional feature; the levels L and V pushed the brake block B against the disk D which rotated with the spindle. At the set speed, the brake would be lightly touching the disk but if the speed increased it pressed harder and would be drawn round with it, against the spring S, so closing the throttle. The balls W can be moved along the arms by a right-and-left-hand screw to var the speed setting which was held to within half a foot per minute even at the highest speed. Speed control was observed on the 'speedle', a water column in an open glass tube near the engine. The head was maintained by a centrifugal pump worked through gearing off the shaft. There was a scale, experimentally calibrated, so that a three and a half inch movement of the water column corresponded to a change of 10 feet/minute in speed when running at 300 feet/minute.

There were three pulleys on the engine shaft and moving the driving strap gave three speed ranges. Adjusting the weights on the governor gave fine control as, for example, on the middle range from 200 to 330 feet/minute.

11.8 Procedures and Calibration Runs

Froude set out his views on operating procedures in December 1870 (ref 11.8).

> 'The experiments will go on at the same time as the model making, the experiments being made at intervals sufficient to enable the water to restore itself completely but often enough to enable a model to be completely tried in a day, for one model a day is the rate at which the moulding plant admits of construction.'

> 'The programme is this – the first thing in the morning the model which was run (ie cast) the day before must be hoisted out of the mould and suspended over it while it is being worked. When clear, the model must travel across to the space that would be occupied if the machine table were in its most westerly position and there the chucking must be fixed. The model must be unveiled and the machine brought under it on which it can be chucked and cut. The mould will then be prepared for the next model. The first model is weighed, run out and lowered into the water where it is ballasted and trimmed. Some time that evening it is destroyed ready for melting.'

This was an ambitious programme but one which Froude was able to achieve. To shape a model in one day and complete the tests the following would be seen as quite impossible today but Froude was using models only six feet long. Cutting time must depend roughly on the surface area of the model and today's models are nearly three times as long or nine times the area.

Calibration runs began on 22nd May 1871 and these enabled operating procedures to be developed and some problems to be overcome. A test in December 1871 was seen as particularly significant, Henry Brunel writing to his elder brother, Isambard, 'They tried their first real experiment in the tank on Saturday, towing a model of GREYHOUND with great success, the results agreeing with the full scale experiment.' it is unclear why that of 3rd March 1872 was officially recorded as 'the first'.

During these early runs it was discovered that it was necessary to wait 20 minutes between runs to allow the waves to die out and for bigger, 600 pound models the interval was increased to

30 minutes. Even with today's bigger tanks, the waves generated by models take about 20 minutes to die down, allowing time for preliminary analysis between runs. Small floats 5/8 inch square and ballasted at one end were used to observe 'quiescence'. Starting in the narrow, shallow end of the tank generated larger waves and was to be avoided. 'Floating curtains' were used every other day to skim scum from the surface.

For the Ramus tests in December 1872 (Chapter 13) a wooden breakwater was installed at the north end with a four foot sluice, 18 inches deep, through which the models passed. This was seen as somewhat hazardous as the distance between the supports of the breakwater was not 'safely sufficient' to allow the chucking beam to pass and, on at least one occasion, there was a collision. The breakwater was not used with paraffin models,

Another problem came to light in December 1872, Henry writing on 27th that 'They've been towing long wooden models in the tank, horrid thing came of it. The tank is virtually 30 feet shorter than it is. It is necessary to start big models in the broad part of the tank for if started at the very end, model makes a great suck of water from which great waves make the model pitch.'

In November 1872 a scheme was devised to accelerate the model speed more quickly, probably also part of the Ramus work. A rope was hooked to the truck leading through fixed blocks about 70 feet from the starting end. The other end was fastened and a block was arranged in the bight which was pulled by several men towards the starting end giving an accelerative force on the truck of half that exerted by the men at double their speed.

Figure 11.9 – This shows the wax melting equipment at Torquay.

Chapter 11: Mr Froude's Tank at Torquay

Figure 11.10 – The models were cast in a clay mould contained in a wooden box, the clay moulded into shape round rough wooden transverse templates giving the desired external shape, with an allowance for cutting. It took about a day to make the mould and cast the model which was ready to cut the following day.

These templates were located in notches cut into the thick, top member of the box. These notches also located similar and smaller templates defining the internal shape of the model. The core of the mould was formed by pinning thin wood laths to the templates at intervals and covering them with calico. This would then be sealed with two skims of soft clay with a film of plaster of Paris between. The buoyancy of the core when the molten wax was poured could be up to 800 pounds so it was ballasted and cold water poured in during the cast.

11.9 Wax Model Making

Froude described his model making machinery in great and loving detail to the Institution of Mechanical Engineers in 1873 (ref 11.9) and it need only be summarised here. Further detail is contained in the extended captions in figs 11.9 to 11.15. He realised from the start that making wooden models, to a high and uniform finish, would be very expensive and he looked for a material which could be cast easily at relatively low temperatures, impervious to water, easy to cut in any direction (no grain) and which would also take a good finish. It should

Figure 11.11 – After casting, the top of the model would be planed flat. This photo was taken at Haslar showing Robert Fuge at work. The procedure at Torquay was very similar.

The Way of a Ship in the Midst of the Sea

Figure 11.12 – This shows the templates which controlled the cutting machine. Each template can be envisaged as a horizontal slice through the model, 5/8 inch deep. The template itself was a thin (1/100 inch), flexible steel strip which, when set, represented a full length, half width plan of the waterline.

The strip was held to the desired shape by a series of square steel adjuster rods which were hinged at clamps through which their length may be altered (seen on the lower shelf). The other leg of the hinge was fastened to a wooden strong back. The template was three feet long, roughly one-third of the length of a model, and the motion of the cutting machine is brought to the correct length by gearing.

If the complete set of templates for a model was laid on top of each other they gave a vivid impression of its shape; Froude sometimes referring to the nest of templates as the 'Model Designer' (upper shelf). It was also possible to run a thin batten round a transverse section to check that it was fair. Adjustments could be made to fair the complete form without redrawing. Models derived from a basic form, say 10% more beam, could be produced from the same template by varying the cutter gear ratio.

Figure 11.13 – The model was fixed to the cutting table, upside down, by pins passing through locating holes in two wooden beams screwed to blocks cast into the wax model. The table was carried on wheels running on rails with grooved wheels on one side only. Longitudinal motion was by an endless cord moved by a hand wheel.

The table with the model would be lifted on four interconnected screw jacks after each waterline had been cut so that the cutters would be in position for the next waterline.

When all waterlines were cut, the model would have a series of steps in the side, the inner corners of which were close to the final shape. Before finishing, a rowel spur was run along this corner making a series of small punctures, alternately almost imperceptible and 'sensible'. These punctures were filled with black lead and, in finishing, the shallower punctures could be taken out but those slightly deeper should remain visible.

156

Chapter 11: Mr Froude's Tank at Torquay

Figure 11.14 – In principle, the operation of the cutting machine was simple; the operator moved his tracer over the template (bottom left) and cutters either side, rotating at 1,500 rpm, cut the waterline to shape. The mechanism required to carry out these actions was quite complicated. For a start, the templates were too thin to bear any pressure and hence the tracer was finely balanced and the operator watched a pointer which told him when contact was made and only just. The shape of the tracer corresponded to that of the cutters, a three inch diameter circle, but because the longitudinal scale of the template was reduced, the tracer had to be an ellipse. This motion was then transmitted to the cutters by a linkage which had to be strong, light and free of vibration needing careful counterbalancing and a large oil filled damper. The cutters were belt driven off the workshop shaft, turned by another steam engine.

The finishing cut could be as thin as 1/1000 inch and these light shavings could be thrown round the workshop. To prevent this, canvas curtains were rigged.

also be fairly strong and not have too high a specific gravity. His first answer was paraffin wax (he spelt it paraffine and sometimes called it stearine). Wax has been used for 100 years or more and has proved very satisfactory. Wax models can easily be altered by cutting off material or, with only a little more difficulty, melting on addition wax. Modern glass reinforced plastic (GRP) models are often started as wax models round which a female mould is laid up in GRP inside which the final model is made.

Froude carried out some experiments on melting wax during November 1870 using steam passed through a coil of pipe, one

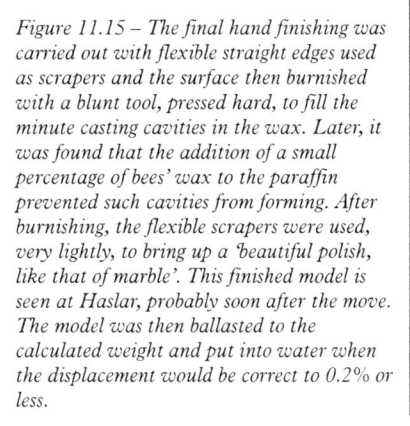

Figure 11.15 – The final hand finishing was carried out with flexible straight edges used as scrapers and the surface then burnished with a blunt tool, pressed hard, to fill the minute casting cavities in the wax. Later, it was found that the addition of a small percentage of bees' wax to the paraffin prevented such cavities from forming. After burnishing, the flexible scrapers were used, very lightly, to bring up a 'beautiful polish, like that of marble'. This finished model is seen at Haslar, probably soon after the move. The model was then ballasted to the calculated weight and put into water when the displacement would be correct to 0.2% or less.

157

and a half inch external circumference and 15 feet long finding that 18 pounds of wax would melt in an hour. A six foot model needed 60 pounds of wax. As experience in testing increased Froude tended to the use of 10 foot models, hollow, with a wall thickness of about one and a half inches when finished and weighing about 200 pounds.

It took about five minutes to cut a waterline; starting the cut from the largest section and running out to one end, then returning to the middle and cutting to the other end. The model cutting machine was ready by May 1871. The rest of the workshop machinery was completed about the same time, mainly from Bodley Bros, Exeter.

Froude had always relied to a considerable extent on graphical integration and Henry Brunel designed a graph paper ruling machine. (Fig 11.16). Ordinary graph paper changes shape with the humidity of the air and is not always accurate. Initially it was worked by hand, ruling first in one direction and then the other, but it was motorised after World War I. It remained in use till about 1960 when the Ordnance Survey were able to supply graph paper of sufficiently high quality until graphical integration was made unnecessary by the computer. Henry's drawings are preserved at Bristol and the machine itself, still in working order, at Haslar. It cost £57-19-4 to make.

Figure 11.16 – The graph paper ruling machine designed by Henry Brunel. It was motorised after the First World War and remained in use until about 1960.

11.10 Staff

William Froude's team at Torquay included several brilliant men, some of whom remained for a considerable time, others gained experience and moved on to win fame elsewhere. Edmund was the senior assistant and ran the tank when William was away. Edmund took part in a few trials and was hurt, not seriously, during the GREYHOUND trial of September 1872. Later that year, in December (ref 11.10), Henry Brunel wrote that he was sorry that Edmund had lost a finger nail and that though a band saw was safer than a circular saw it was so much nicer to use that it was used more and hence not so safe. 'The only really safe way of using a band saw is on 'the borrow a watch principle' – always get someone else to use it when you can.' There were other health problems; Beauchamp Tower was ill in June 1872 and William retired to bed with gout in December 1873.

When the tank was ready, in December 1872, the Admiralty appointed an additional assistant, Phillip Watts. He was a new graduate from the Royal School of Naval Architecture at South Kensington and Froude was very pleased with him. He told the Royal Commission of 1872 (ref 11.11) 'I have an assistant who was educated there who has a very competent knowledge indeed of the science of naval architecture.' And later in his evidence said of the School and Watts 'I do not know much of its operation. I only know of its fruits, in the fact that a gentleman who is now acting as one of my assistants in my investigations, and who was educated there has evidently been extremely well taught and is not only well up in the mathematics of naval architecture but in the common sense of the questions dealt with...'

Until well after World War II the Admiralty continue to send some assistant constructors (including the author) to the 'Admiralty Experiment Works', a title taken by the tank from 1877. Watts was not at Torquay for very long, his relief Perret was appointed in March 1873. Watts was still there in August when Henry Brunel wrote that Watts had told him that the Constructors were very much impressed by Eddy during their visit (ref 11.12). Watts later left the Admiralty to become chief naval architect with Armstrongs, returning to the Admiralty in 1902 as Director of Naval Construction and was then responsible for the design of most of the Grand Fleet of World War I.

J. R. Perret graduated from the Royal School in May 1871. In December 1879 his pay at Torquay as a 'Draughtsman' was £220 gross, less tax of £2-1-8, paid by the Admiralty. Perret later became Technical Director of Armstrongs. Another Royal School man was F. P. Purvis. In October 1872 Henry Brunel was teaching both Perret and Purvis to use 'pens', presumably the graph paper ruling machine. Purvis was paid less than Perret and soon left to complete and run the Denny Tank later becoming the first Japanese Professor of Naval Architecture at Nagasaki.

In an interesting letter of October 1873 (ref 11.13). Henry suggests that the output from Torquay was limited by the insufficient number of graduates.

> 'Although enough has been done to make the Admiralty almost foolishly enthusiastic, the experiments are not going as they ought, the utmost use isn't made of the tank and experiments are interrupted, not for want of appliance or workmen but because there is not enough skilled brainpower on the establishment to keep it all going. This needs three or four Eddys – beside Eddy wants a holiday – and it would be useful if Froude and Eddy were able to go away without work coming to a standstill. It isn't that mere assistance wanted at Eddy's level but other things have to stand still and a whole bunch of subjects want taking up by someone and Froude is fussed at the non-execution of work. Froude should ask the Admiralty for more money to employ more assistants to get on quicker with the work. Hurrell has mathematical knowledge and would tackle work better than Eddy. Eddy gets £150 and he's getting old. Froude said that he must get permission to raise his pay. Hurrell should marry and settle in Torquay.'

Hurrell had returned from India in 1871 with about £5,000 of savings to invest. Both his family and Henry were worried, the latter writing (ref 11.14) 'It is no injustice to say that Hurrell is willing to be the partner of a plausible scamp, he has already contemplated doing so, great care is needed.' Henry also pointed out that his £5,000 at 6% would give him £300 and thought that he was worth £600 as William's principal assistant. There were somewhat similar hopes and fears over Hurrell's marriage: hopes that he would marry and fear that he would marry the wrong girl. Eventually in 1874, there was a hasty note to Henry from Hurrell asking him to go down into his room and destroy the letters from Agnes Wilberforce as he was to be engaged to Beatrice Ryder. This marriage seems to have been successful but Beatrice died and, in 1881, Hurrell did finally marry Agnes. His business partnership also prospered with Heenan and also with Arthur Lucas in the manufacture of Tower's 'Spherical Engine'.

Pay records exist (ref 11.15) from which one can identify the rest of Froudes's staff and they all seem to have stayed with him for many years. The statement for April 1876 is reproduced as typical:

Chapter 11: Mr Froude's Tank at Torquay

R. E. Froude............£150 (+ free board and lodging?]
                                    £ - s - d per week
F. P. Purvis 3-10- 0
J. R. Perrett 4- 4- 0
C. Manning 1- 4- 0
also:
W. Ackrell 1- 4- 0
G. Kingdom 2- 2- 0
J. Toby 1-16- 0
J. Bond (apprentice) 1-10- 0
G. Turpin 1- 5- 0
S. Hutchings...................... 1- 5- 0
J. Tippett 14- 0
S. Jerman 1- 4- 0
G. Pugsley 1- 2- 0
G. Pugsley 1- 2- 0

Note: Manning is grouped with Purvis and Perrett but separated by the word 'also', there were two G. Pugsleys.

Arnulph Mallock was working with Froude in 1871–1872 but he had only just graduated and was essentially still training. His day was yet to come.

11.11 Finance

It is always difficult to estimate the likely cost of novel research and the novelty of Froude's work made it even more difficult. The accounts were inevitably difficult as work was going on the tank itself, paid out of the Admiralty's grant of £2,000, on experiments for the Committee on Designs and on specific trials, paid for on an individual basis and Froude's staff were also working on the house and garden. Surviving time sheets show that the man hours actually worked by the staff were carefully recorded and charged; the difficulty lay in estimating the hours which would be needed. Henry Brunel had spent some time in showing Eddy how to keep the account books.

The main work on the tank in the first two year was on a 'fixed price' basis and there was no requirement to produce accounts to the Admiralty. However it was decided, at Henry's suggestion, to keep a 'General Record Book' for 'inspection by exalted individuals' and happily, this survives today (ref 11.16). The building of the tank cost much more than Froude's estimate and in May 1872 he told a Royal Commission that he had already spent more than the £2,000 allocated even though work had hardly started. 'I had hoped it would prove sufficient but I find it is not and I have just received an intimation from the Admiralty that they are prepared to extend the grant.' (ref 11.17)

He was asked if his own services were remunerated and

replied 'No – a son of mine, who is clever and a good mechanic, and understands the question thoroughly, is employed by me as an assistant and he is paid £150 per annum. That I put into my programme as part of what I proposed, and therefore, I have no scruple in adopting that arrangement, but my own services are quite gratuitous; I am, however, thoroughly glad to give them, the matter is of such extreme interest to me. It is a great demand on my time, but I am very glad to give the time.' With hindsight, it may be seen that the estimate with his 1868 proposal included virtually nothing for the pay of his staff during the building of the tank and a quite inadequate amount for the two following years.

Froude explained that he had made considerable savings by making his own apparatus in a rough yet effective way, where, if a regular manufacturer had been applied to it, it would have assumed an expensive character.' He pointed out that many great men, such as Sir William Thompson, were skilled in the making of inexpensive apparatus.

The admiralty seems to have gone out of its way to help by making additional payment, by paying directly the salaries of Watts and later Perret and on 1st August 1876 the annual payment was raised to £1,550. This sum still seems rather small bearing in mind that by 1876 the annual salary bill came to £1,330, including £150 for Edmund. Some idea of the difficulty in forecasting expenditure is obtained by examining the quarterly claims for 1877–1878.

	£ - s - d
June 1877	420-15- 6
Sept 1877	356-12-11
Dec 1877	163-16-11
March 1878	610- 8-11

Quarterly variations of almost four to one are indeed hard to predict though one may wonder why the December figure was less than the salary bill for the period. (ca £325)

The Admiralty accountant continued to help, allowing advances on quarterly claims, and advising how to distribute such claims to fit best within each year's budget. Trials such as those of the SHAH and IRIS were funded separately from the annual grant, the bill for IRIS coming to £108-16-0 including Henry Brunel's fee of £21, the rest being expenses. (This implies that the salaries of those taking part were still charged to the annual grant.) By 1877 there was Froude's very considerable involvement with the work of the INFLEXIBLE committee further to complicate matters. It is not surprising that by the time of William Froude's death in 1879 the accounts were in a muddle.

The Accountant General of the Navy wrote to Edmund during Williams's visit to South Africa on 25th February 1879 'If desired, the bill made out to William Froude on 28th January and 14th February should be paid on your endorsement, they should be returned to this department for alteration. I should be glad of the production of any written authority you may possess to sign for Mr. Froude.'

There followed a long investigation of the accounts, initially it seems with some suspicion. The Accountant General wrote a long letter on 6th April 1880 which can be summarised as follows. 'My Lords have your letter of 17th ultimo with accompanying statement relating to the claim of that William Froude against the Admiralty in connection with towing and rolling experiments conducted during several years from 1871 onwards.'

> 'It appears from these statements that Mr. Froude incurred considerable expenses of which no repayment has apparently been made.' –
>
> 'Their Lordships consider the conditions hardly satisfied by the documents forwarded as other work for the Admiralty was being done for which payment was made.' –

The Accountant then proposed to send Mr. Bather to Torquay to enquire and report. This report was presumably favourable as on 11th June 1880 the Admiralty forwarded an order for £1673-1-10 outstanding, subject to production of a receipt from Mr. Brunel for the amount due to him.

To build a novel research laboratory, get its equipment working and prove the procedures between February 1870 and December 1871 was a notable achievement. Froude acknowledged that it was a team effort (ref 11.18).

> 'The great success of the machinery employed was the result of the combined thought of several minds, whereby the system had been matured into its present very perfect state, enabling the shaping of the models to be done with very great accuracy.'

None of those involved doubted that the inspiration and leadership was that of William Froude.

11.12   Froude on Committees

Froude served on a number of Government Committees and others such as those of the British Association, and his work for them has been described. However, on 29th May 1872 he gave lengthy evidence to the Duke of Devonshire's Royal Commission on the Advancement of Science (ref 11.19). While much of his evidence was in connection with particular investigations he also gave some interesting views on the involvement of Government in Science which are worth introducing here. Since Froude was answering specific questions, there was considerable duplication and the following passages, whilst following him closely, have been considerably edited.

Froude thought that it was the duty of government to assist in scientific advance. It seemed an object of national credit and national importance that scientific knowledge in the country should be advanced in both basic and applied aspects to the highest degree and that the nation valued such advancement and would approve its promotion by the government.

His own work had been chiefly in one department where scientific knowledge was of special monetary value to the government. He gave his views on where such knowledge would be of most value. 'In questions of guns and gunnery, and of naval architecture, in the construction of public works generally, it is of great pecuniary importance to the government that the most scientific methods should be pursued; and though these are more generally understood than formerly, the assistance of the highest scientific knowledge in the country would have been and would now be of immense pecuniary value.'

It was less easy to set out the steps which should be taken but Froude thought that there should be some method in which the highest scientific knowledge of the country could be made available and made effective in all departments of science which government ought to promote. At that time there was no direct scientific input and it was only found at rare intervals when the government called on one or two individuals rather than as a 'constitutional assemblage' of the highest knowledge.

Colonel Strong had proposed a 'Scientific Council' to advise the Government and Froude was invited to comment. He clearly had difficulty in understanding exactly what was proposed and the minutes do not make it clear to a modern reader. Strong's proposal seems to have been for a single Scientific Council with every branch of science represented by at least one member and he thought that there should be other members from the navy and army. Froude saw great difficulty in making sure that the best men in each branch were chosen and said:

> 'It seems the essence of the thing that the very best knowledge that the country possesses should be available in the particular department to which it is applicable. I think, however, that such a body, if selected in the best manner, would know when to act and how to act; and that it would be able to impress on the government and to impress on the department its own views'

He pointed out that the legal profession did have a high consultative body and that science should be on the same footing.

In government departments where scientific knowledge is available and necessary the head man should be possessed of a very high class of scientific knowledge but there would still be occasions when he would be glad to draw on a more specialised knowledge. for example, the Chief Constructor of the Navy should have a general and proper knowledge of the scientific

Chapter 11: Mr Froude's Tank at Torquay

principles which his department requires but, on extraordinary occasions, he ought to be able to draw on superior knowledge applicable to his department. For this reason a council would be of use but its existence would not relieve the government of the necessity of putting a very capable man in every department in which scientific knowledge is required.

The Commission suggested that Strong's council would have only two or three men with any knowledge of ships. Froude responded at length saying that a man of high mathematical attainments would inevitably know a great deal of the scientific principles directly bearing on questions of naval architecture. Without being conversant with the actual construction of ships, he would have seen certain principles which ought to have been attended to, but which are only quite lately becoming attended to.

> 'Such a council would make it its business to feel its way as to the particular questions in reference to which its actions would be of value, and would understand how to apply its knowledge. I think the council would have to act with discretion to the departments because I found myself, for instance, last year when I was sitting on the Committee of Design of Ships of War; we were a large and somewhat miscellaneous committee, and a great many questions were asked by members of the Committee which had to be answered by the Constructors Department, and that department was worried out of its life almost by constant demands on its time in preparing reports on various questions, and many questions, I am satisfied, were questions which need not have been answered;.....'

It was a large committee with a great deal of relevant experience and generally a good committee but such a committee 'is very likely to wear out the department by demands on its time, when in many cases they might be better employed.' The problem with such committees was that they took so long to understand the problem and the background or, as Froude put it "the dream and interpretation". This problem with ad hoc committees might be less with a standing committee.

In a letter to Froude (ref 11.20), Henry agreed with Froude's views but was less certain of the wisdom of stating them publicly.

> 'The admiralty might grumble if anything implied that they didn't support him enough; didn't like all his evidence before the Commission – all true but too true regarding different to dragging evidence before the light of day, stirring up matter,...'

The commission was told that the Committee on Designs had made some specific recommendations which Froude still had in hand on the oscillations of ships in a seaway. Froude acknowledged Admiralty support in this work but continued: 'I should have much felt the value of such a Council, had it

165

existed, because I have been in rather an irresponsible position with regards to those experiments. I might have put the Admiralty to very great expense, or, from fear of running to too great an expense, I may be acting too timidly in particular cases.'

He made a similar point in connection with his resistance work: 'And here again, as the conduct of the experiments is practically left to my discretion, I am in a sense irresponsible, and I should feel the value of having men of great eminence to consult. For example, I have found it of great value to have even casual and occasional discussions on the subject with Sir William Thompson and with Professor Rankine. Many new lines of thought are at once suggested by men of that stamp.'

In this passage one may recognise Froude's "sacred duty to doubt"; in this case doubt of his own conclusions. One may also wonder if the proposed "Scientific Council", probably with much the composition of the British Association Committees and opposed to model testing would not have vetoed the tank. Froude said that he reported formally to Dr. Wooley and the Constructor's Department but that full supervision of his work was not possible as it takes as long as doing the job.

'I believe that under present circumstances his [the Chief Constructor's] time must be so very fully occupied by his ordinary work that it can hardly be possible for him to spare time for the scientific interest of new and difficult questions. The introduction of armour plating and very heavy artillery has not involved only very great novelty and difficulty of constructive design, thus enlarging the ordinary labours of the office, but has also rendered necessary the solution of questions, in the treatment of which advanced science has been specially required.'

'I would like to say something in reference to the view that Government is bound to promote scientific investment for the improvement of its own particular work, yet in reference to these investigations in which success, if attained, would be highly remunerative, the prospect of success will secure the investigation. This no doubt often happens, but in such cases the first ventures are often ruinous, and this acts as a deterrent. In many cases, only firms of the highest standing could make the ventures and such firms, already possessing an established lucrative business, often hesitate to incur the danger of loss; perhaps they are even less willing to incur the discredit of failure.'

Asked if he had anything else to say, he added:

'I am not aware of any; I think that one of the things which one has to learn is to know what are the limits of what one knows. I always feel that it is very unwise to step out of the results of one's immediate experience.'

11.13 On the Constructor's Department

'It has certainly seemed to me to be overworked. The mass of work that has to be got through is extremely heavy and there is, besides, an immense deal of paper work and correspondence, and returns and reports on various questions are always being asked for; but I gathered this impression at a period when the Committee [of Designs] was adding greatly to the latter sort of difficulty; still, it seems to me that the Constructor's Department is so heavily occupied as to prevent any of its members from personally entering on any new investigation which involves time and labour, and unembarrassed thought.'

Should they be less harassed with reports etc? – 'Just so'

References

11.1   Held in the Froude's Museum, Haslar.
11.2   Brunel Collection, 11129 090471
11.3   Froudes' Museum, address book.
11.4   Brunel Collection, 110771.
11.5   J. M. Dirzwager. *Contribution of Dr Tideman to the Development of Modern Shipbuilding*. National Maritime Museum 1981.
11.6   Models were designated by two letters from AA, AB etc onwards. In post war years they reached three letters.
11.7   Froudes' Museum, 13th October 1870.
11.8   Froudes' Museum, Note book 1.
11.9   W. Froude. *Description of a machine for shaping models used in experiments on the forms of ships*. Trans. I. Mech. E., London, 1873.
11.10 Brunel collection, 13201 161272.
11.11 Royal Commission on Scientific Instruction and the Advancement of Science. Froude's evidence, 29th May 1872. Parliamentary Papers, 1874, XXII, pages 247-250.
11.12 Brunel collection, 14140 130873.
11.13 Brunel collection, 14333 271073.
11.14 Brunel collection, 11314 71.
11.15 Froudes' Museum.
11.16 Froudes' Museum, 'General Record'.
11.17 As 11.11.
11.18 As 11.9.
11.19 As 11.11.
11.20 Brunel Collection. 010275

The Way of a Ship in the Midst of the Sea

Chapter 12
The Greyhound Trial 1872

12.1   Introduction

In August 1872 William Froude carried out a trial in which the sloop GREYHOUND (fig 12.1) was towed behind another ship and its resistance measured over a range of speeds and in different conditions. This trial was initiated by the Committee on Designs for Ships of War and differed in its aims from that proposed by the British Association even though Bidder and Froude were members of both Committees. Though this trial had no direct connection with the rolling trial of August 1872 also involving the GREYHOUND, some preparations overlapped and some accounts are unclear as to which trial is being discussed.

Figure 12.1 – The only known photo of HMS GREYHOUND, used in rolling experiments and towed to verify Froude's Law.

In the towing trial it was intended initially to measure the resistance of a typical ship over a range of speeds and hence to identify how much of the power developed in the machinery went into overcoming the resistance of the ship. Much of the power went in overcoming friction in the engine and shafting, more in driving auxiliaries such as the air pump and still more in hydrodynamic losses at the propeller. It was rightly suspected that the 'effective' power, overcoming resistance, was only a small proportion of that developed by the engine.

As is usual, and quite proper, such a trial acquired a number of subsidiary objectives, such as measuring the resistance of bilge keels, testing Rankine's formula for resistance and, for Froude, to validate his model test procedure, the latter becoming the primary object. Despite these additions, the purposes of the trial were clearly defined, unlike those proposed by Merrifield whose objective may be described as to see what happened. Froude wrote late: (ref 12.1)

> 'The scale which has been propounded possesses undoubted prima facie theoretical truth, and some experimental justification, and would be tested completely, and might receive correction by help of the trial of a full-sized ship.'

12.2 Experimental Method

It may seem easy to tow a ship and, by measuring the force in the tow rope, obtain the resistance. The author has been involved in planning such a trial himself and knows that it is far from easy even with modern equipment and the experience derived from previous, similar trials. For a start, if the towed ship is too close to, or in the wake of the towing ship, its true speed through the water will be affected by the disturbance created behind the towing ship. Froude arranged a strong boom from the side of the towing ship, ACTIVE so that GREYHOUND was towed 45 feet to one side and 190 feet behind the stern of ACTIVE. A similar precaution involved towing the speed measuring log well clear of the GREYHOUND. It is also necessary that the ships hold a steady course, clear of irregular currents and, in this respect, Froude paid tribute to the skill of the pilot in charge.

The tow rope dipped into the water under its own weight so that the force which it applied was not the horizontal resistance which was required. Froude – the railway engineer – connected the tow rope to a truck running on a short railway on the forecastle of the GREYHOUND and the dynamometer connected to this truck. There was also an arrangement of levers to disconnect the dynamometer from the tow rope when sudden loads came on due to manoeuvring.

The dynamometer differed from those which Froude had used previously in using hydraulic operation. The pull was taken on a piston fitting closely in a 14 inch cylinder from which oil could flow into a much smaller cylinder, one and one eighth of an inch in diameter. The movement of the piston of the small cylinder against a spring was recorded on a rotating drum to give the horizontal force on the ship. The dynamometer was made by Kittoe and Brotherhood and fitted so exactly that there was no leakage of oil even though packing had been omitted to limit friction effects. With the cylinder empty, a force of only 30 pounds was needed to move the piston, negligible in relation to the force being measured. There were still occasional problems with the piston sticking and, in his report to the Admiralty, Froude referred to the need for 'gentle percussive treatment.' In

Chapter 12: The GREYHOUND Trial 1872

his later paper to the INA this was explained as 'a succession of sharp blows with a mallet to the cylinder.'

Bidder suggested that to measure the speed, a 'log ship' (effectively a sea anchor) be thrown overboard; a roll of twine attached to it drove then the drum on which resistance was recorded. A clock added time marks. The log ship moved slowly through the water and a correction, measured at three tenths of a knot, was made for this. A further check that speed was constant was made by noting the rotational speed of ACTIVE's propeller. Froude noted that in any future trial of this sort he would use not a log ship but a screw log held away from the ship. In the discussion, he rejected the use of a pilot tube as too much affected by the flow round the ship.

Froude took many precautions to ensure accuracy, including the use of a gauge to measure wind speed during each run. The effect of wind was measured by letting the ship drift before a 15 knot breeze and measuring the speed. Since the resistance at that speed was known to be 330 pounds, this was the magnitude of the wind force at 15 knots and was taken to vary as the square for other speeds.

Planning involved many individuals and Admiralty departments, Froude returning from one meeting remarked to Edmund that he had fought with Beasts at Ephesus.' (ref 12.2) Froude acknowledged the help of Bidder in planning the trial and in the design of the apparatus. The preparation of GREYHOUND, including removal of the masts to reduce wind drag, was in the hands of Henry Brunel. The trials were conducted by Froude's new assistant Constructor, Phillip Watts, Henry writing (ref 12.3) 'Watts hasn't been treated well. The Admiralty have chiselled him out of some of his expenses...'

12.3 The Trials

The GREYHOUND was towed at three displacements and at a number of trims. At each condition her resistance was measured over a range of speeds from three to twelve and a half knots. Finally, bilge keels 3ft 6in deep and 100 feet long were added and the resistance re-measured at normal displacement and three trims. All trials took place in Spithead to ensure calm water and occupied six weeks except for a few days interruption by bad weather When a ship accelerates or decelerates, the mass whose speed is to be changed is not just that of the ship but includes that of the water dragged along by the ship, known as 'virtual mass'. Froude studied this effect in trials in which the tow was dropped at 12 knots and the rate at which GREYHOUND slowed was measured. There were also some runs in which she was accelerated by the power of ACTIVE but these could not start from rest as the snatch in the tow would have been too great. He found that the total mass involved was about 20% greater than that of the ship alone. A small correction was derived from this result and applied to runs where there had been a change during the recording.

12.4 Results

Up to 8 knots the resistance varied as the square of the speed, equal to $88v^2$ pounds where v is the speed in Knots but rising more rapidly at higher speeds

Speed Knots	Resistance Tons	$88v^2/2240$
4	0.6	0.6
6	1.4	1.4
8	2.5	2.5
10	4.7	3.9
12	9.0	5.7

Perhaps many readers will be surprised that the resistance of a 1,000 tonne ship is so small but that is why sea transport at low and moderate speeds is cheap.

The effect of changing draught was interesting; the associated change in power was not proportional to either the midships area or to Displacement to the two thirds power showing the limitations of the Admiralty Coefficient (Chapter 8). It also suggested to Froude that a deeper ship might need relatively less power than a beamy one. He followed this up with models but found that the situation was too complicated to reduce to such a simple rule.

The effect of trim was not great (without bilge keels), trim by the stern was beneficial at higher speeds by up to 7–8% reduced resistance. With bilge keels, there was no advantage in stern trim. The additional drag of bilge keels was much less than expected; at 10 knots the frictional resistance of the bilge keels should have been 800 pounds whereas the experimental results appeared to show only 300 pounds. Froude suggested that the condition of GREYHOUND's bottom might have altered during the week she was in dock while the keels were fitted and this seems a very likely explanation. The actual water speed over the bilge keels, close to the hull, would also have been less than the speed of the ship as she would draw water along with her. The lower resistance, combined with the benefits shown in the rolling trials (Chapter 5), made a convincing case for large bilge keels, not fully recognised at the time or even today.

Froude then compared the power developed within GREYHOUND's engine during a self propelled trial and the 'effective' power corresponding to the resistance of the hull at the same speed, finding that, at best, there was a loss of 58% He also tried to compare effective power with that corresponding to the thrust developed by the propeller.

Chapter 12: The GREYHOUND Trial 1872

'Making the utmost allowance for engine friction, etc, it seems from this impossible to doubt that the actual thrust delivered by the screw-shaft is largely in excess of the resistance due to the ship, and that considerable extra resistance must be caused to the ship by the action of the screw, by the diminution which that action causes on the hydrostatic (or perhaps I should say hydrodynamic) pressure of the water against the contiguous parts of her run. I have often insisted on this effect of the screw working in 'dead-water' close to the ship's stern – although, considering the great importance of the subject, I have had but small success in my endeavours to draw attention to it, and I believe that there is a general supposition that the effect is but small in the case of a ship having as fairly fine a run as the GREYHOUND. Nevertheless the above comparison appears to show that it is considerable; and this is corroborated by six of the towing-experiments at "normal trim and displacement" taken with the screw lowered into its working position. In three of these six experiments, the screw was allowed to revolve; and it will be seen that in these three cases the resistance of the ship is much higher that at the same speed with the screw lifted, nay even with it down and fixed.'

To Froude and to modern naval architects the real interest is in the comparison between the resistance of GREYHOUND as measured during the trial and that estimated by scaling up the resistance of the model. Fig 12.2 clearly shows that curve A deduced from a varnished model is parallel at all speeds to curve B, that of the ship. Froude pointed out that the comparison with a varnished model was for convenience only; it was a surface which could be reproduced easily and consistently. Using a model "surface" corresponding to two thirds varnish and one third calico, curve C, agreement was almost precise.

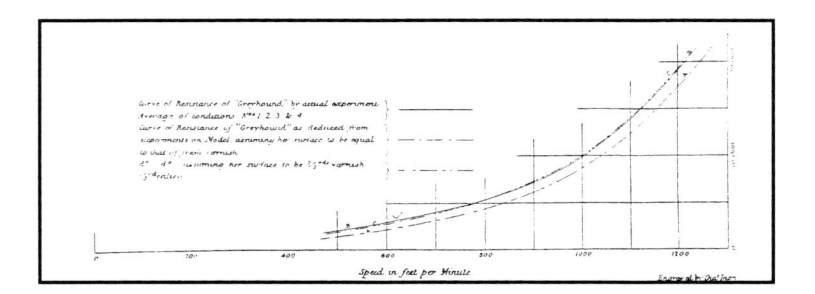

Figure 12.2 – Actual resistance of Ship compared with that above deduced from Model.

Froude was probably aware that the effect of surface roughness is more complicated than this and in the next chapter his ideas will be outlined. He certainly realised that the copper sheathing on the bottom of the GREYHOUND was in poor condition which would add to the frictional resistance. It is also quite likely that some fouling would develop in the copper, which was in poor condition, within a day of undocking, adding considerably to the resistance, as happened during a trial with similar objectives after World War II (ref 12.4).

Froude observed the similarity of the waves generated by model and ship at corresponding speeds, further confirming his law by comparison. Phillip Watts was lowered over the bow to sketch the ship's wave formation but, unfortunately, these records have been lost. Froude said during the discussion of his paper:

> 'In watching these waves one is struck, if one has seen the trials of the ship and the model experiments, with the resemblance between the waves created by the model and those created by the ship – at what I call the corresponding speed'.

Froude reported his work to Nathaniel Barnaby (Reed's successor as head of the design department), an outstanding theoretical naval architect, who seems to have accepted that the trial had fully validated Froude's method of model testing. The tone of the discussion of Froude's paper on GREYHOUND at the INA was a complete contrast to the hostility shown during the discussion of Merrifield's paper in 1869 (Chapter 10). All concerned now accepted Froude's views; indeed it would seem that the only one with doubts was Froude himself. At the time, Froude's conclusions were firm, saying that:

> 'The experiments with the ship, when compared with those tried with her model, substantially verify the law of comparison which has been propounded by me as governing the relation between resistance of ships and their models. This justifies the reliance I have placed on the method of investigating the effects of variation of form with varied models – a method which, if trustworthy, is equally trustworthy for testing abstract formulae, or for feeling the way towards perfection by a strictly inductive process.'

He was very dismissive of Rankine's formula saying that as he had not been able to construct any general expression which was theoretically acceptable for estimating the resistance of a ship, he could not comment on the validity of other formulae, a view fully accepted today. The editors of "Naval Science" (Reed and Wooley) in reporting the GREYHOUND trials put checks on Rankine's formula. Since it only purported to estimate frictional resistance it could only be used to estimate GREYHOUND's at low speed where it was considerably in error. They also suggested that it was wrong in principle.

By 1874 Froude had doubts over two aspects of the trial, the possibility that the towing ship was too close and the effect of the shallow water in which the trials were run. The possibility of interference had occurred to Froude while developing his propeller dynamometer when some tests showed him that mutual interference between two bodies close together in the water could be important. He then ran tests with the

GREYHOUND model in its usual place under the carriage with another model (actually that of the SHAH) in the relative position of ACTIVE, to scale. He found that this had no effect on the resistance of the GREYHOUND and that it was not necessary to make a special model of ACTIVE. Once again, one must admire Froude's powers of observation and his lateral thinking which led him to read across from calibrating a propeller dynamometer to interference between ships (ref 12.5).

The effect of shallow water proved to be more serious. He made a false bottom in the tank covering 60 feet of the length which could be arranged at between 2 feet (corresponding to 7 fathoms for the ship) and 6 feet below the surface. He found that the model resistance increased suddenly as it entered shallow water and then fell to a level some 5% below that in deep water. the results were interpreted as showing that the estimated resistance from the varnished model should be increased by 6.5% at 6.8 knots, ship speed, and by 4.5% at 11.75 knots. He then pointed out that resistance of big ships on the Stokes' Bay measured mile with a depth of 12 fathoms would be increased by about 10%.

Confirmation of his views and the overthrow of the opposition must have given Froude great satisfaction and to this day his success is commemorated in Froude's Law (of comparison) and the Froude Number to define his "corresponding speed". There are a few more errors which he was not aware of: the model of the GREYHOUND was too small for really accurate results and there would have been some laminar flow at the bow. It is most likely that she also had a significant amount of 'viscous pressure drag' which does not follow his Law.

In particular, Froude wrote:
> '...my experience in the conduct of these experiments fully bore out the views which I had previously expressed, of the almost impossibility of entering a comprehensive investigation of the properties of different forms of ship by full-sized towing-trials.'

Instead, he used the GREYHOUND form, now so well documented, as the starting point for a series of model tests of variants of this form. Froude did not attach great importance to the GREYHOUND trial as to him it merely added confirmation to what he already believed, but to most naval architects it was the essential proof of his law of comparison.

References

12.1 W. Froude. *On experiments with HMS GREYHOUND*. Trans. INA., London, 1874. Care is needed to distinguish between this trial and Froude's rolling trial using the GREYHOUND in August–September 1872. (Chapter 6)

12.2 Professor Sir Westcott Abel. *William Froude – His Life and Work*. Devonshire Association, 1933.

12.3 Brunel Collection. 130873

12.4 J. F. C. Conn, H. Lackenby & W. P. Walker. *BSRA Resistance Experiments on the LUCY ASHTON. Part II. The ship-model correlation for the naked hull condition*. Trans. INA., 1953.

12.5 R. W. L. Gawn. *Historical notes on Investigations at the Admiralty Experiment Works, Torquay*. Trans. INA., London, 1941

Chapter 13
Years of Achievement 1872–79

13.1   Introduction

The SWAN & RAVEN tests had confirmed that Froude's Law of Comparison was basically correct but there was still much to be done. In scaling model results to ship size, frictional resistance had to be estimated and separated from the rest but there were no frictional data available and hence the first experiments in the new tank were to obtain this information. There were some loose ends in the definition of residuary resistance and the action of waves and eddies had not been completely described.

The GREYHOUND trial had also confirmed Froude's long held views on the importance of the propeller and its interaction with the flow round the hull and this opened a another novel field of investigation requiring the design of new apparatus and the development of experimental procedures. In turn, these led to more precise definition of the parameters involved in propeller performance, a process completed by Edmund Froude after his father's death. Speed trials carried out using Denny's methodical approach, explained later, suggested further studies.

These advances in theory, experimental methods and instrumentation could not be divorced from the major task of developing new and improved hull forms for ships of the Royal Navy and very often an apparently routine test would lead to advances in theory and vice versa. For clarity, it is necessary to follow each topic to a conclusion even though it may be necessary to return later to consider a different aspect of the same experiment.

The sections which follow will consider in turn:
  Resistance components, friction, wave making and eddy making.
  General hull form.
  Propeller performance and interaction.
  Trials.
  Tests on specific forms.

Froude's work came at a time of great changes in warships. Reed had directed or sponsored new approaches to the arrangement and protection of warships, to the design of their structure and to their stability while higher steam pressures, used more efficiently in compound engines, produced the power needed for higher speeds while their greater efficiency made it possible to give up sail in most ships. Hull forms and propellers which would satisfy these somewhat conflicting requirements could be designed with confidence using the results of model tests.

13.2 Part I – Resistance Components

13.2.1 Friction

13.2.1.1 Lessons from the Torquay Water Main

Since Beaufoy's work (Chapter 8) there had been an increasing awareness of the importance of frictional resistance; Scott Russell used a formula based on Beaufoy's work. Froude's work in measuring the pressure drop along the Torquay water main and the subsequent scraping of the pipe gave him a new insight into friction which he put forward in a short paper to the British Association in 1869 (ref 13.1). The paper began by pointing out the errors in the contemporary views of friction. The most obvious was that what he called the 'quality of surface' (roughness) was completely ignored as it was assumed, incorrectly, that water would 'fill in' the roughness so that the rest of the flow would slide over a smooth surface.

He also showed that the formula then in use which stated that flow in a pipe was proportional to speed to the power 2.5 led to impossible conclusions. The view that the last square foot of a flat surface experienced the same drag as an earlier square foot ignored the obvious point that the relative speed of water over a surface would be reduced by friction so that the downstream sections would be exposed to a lower relative velocity.

Froude then put forward a number of propositions, expressed with rather less than his usual clarity, which, simplified a little, were that the fluid will not slide bodily over the surface as with two solids sliding but, in the layer of the fluid next to the surface, there will be a gradient of velocity from particles at rest on the surface to that of the free stream only a little distance away. This implied that a truly smooth surface, wetted by the fluid, will experience the same resistance as if it consisted of the fluid and that the body will give momentum to the fluid.

These propositions amount to a simple statement of what is now called boundary layer theory. In a later paper (ref 13.2) Froude acknowledged the value of discussions and correspondence with the then leaders in that theory, Professor Rankine, Sir William Thompson and Professor Stokes. Froude's contribution was to apply the basic theory to the solution of very practical problems.

13.2.1.2 "Plank" Tests

Froude modified his experimental and scaling procedures in detail from time to time but the outline which follows is a general representation of his methods for scaling from model results to ship values.

(a) Measure the total resistance of the model (R_{tm}) over a range of speeds.
(b) Subtract the frictional resistance of the model (R_{fm})

Chapter 13: Years of Achievement 1872–79

from the total (R_{tm}) to give model residuary resistance. (R_{rm})
(c) Scale model residuary to ship value (R_{rs}) by the law of comparison for the corresponding speed at which the resistance per unit displacement is the same.
(d) Add ship frictional resistance (R_{fs}) to give ship total resistance at each speed. (R_{ts})

However, there was little or no data on the values to be used for model and ship frictional resistance and hence one of the early tasks of the tank at Torquay was to provide tabulated data. A number of flat surfaces of different lengths – planks – were towed down the tank under the carriage over a range of speeds. The planks were 3/16 inch thick and 19 inches deep, top to bottom, arranged with the top edge 1.5 inch below the water surface. A lead keel of the same thickness as the plank was fitted to give neutral buoyancy,

It was important that the plank was held precisely in line with the tank axis during the experiments and to ensure that this was so the plank was fixed to a light but strong wooden girder. This was attached by a rocking frame to the carriage so that it was free to move backwards and forwards but was rigid against sideways motion.

The task was to obtain data on the effect on frictional resistance of speed, of varying length in the direction of motion and of surface roughness. For convenience, each plank was tested over the full range of speed and graphs plotted of resistance against speed.

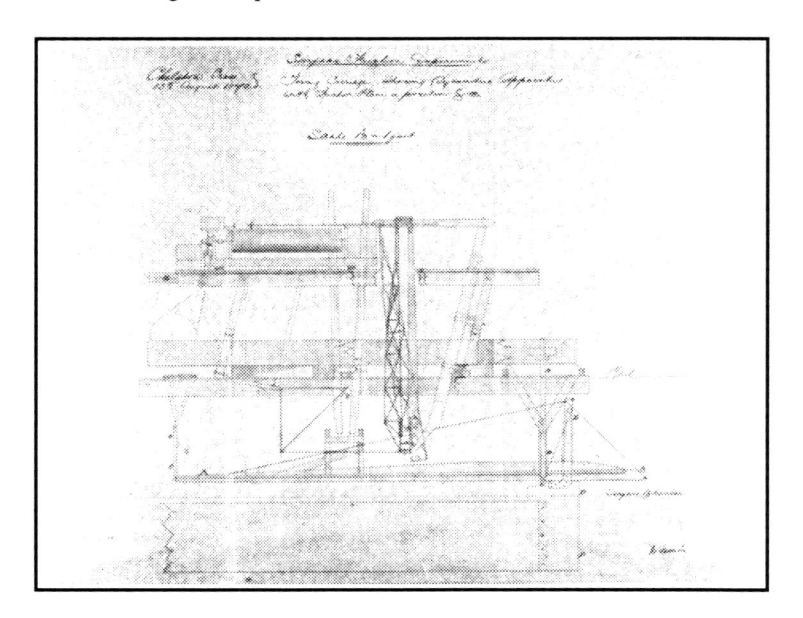

Figure 13.1 – Froude's drawing of the arrangements to measure the frictional resistance of flat planks. (His signature is bottom right.)

Froude sent a preliminary report to the Admiralty on 10th August 1872 which described the tank, the apparatus and the methods used (ref 13.3). Preliminary results and analysis were also included. The plottings of resistance against speed showed that resistance varied at a power of speed of about 1.8. The corresponding plots of resistance against length at constant speed revealed a number of experimental problems which took some time and thought to solve.

It was clear that the resistance per square foot diminished as length increased and that this variation was greatest in the first two feet. At these shorter lengths it was not possible to draw smooth curves through the experimental results. The tests were repeated, with even greater care, but with the same result.

Froude then tried a much sharper leading edge or cutwater which improved the lie of the data a little. The square trailing edge was then re-shaped with a fine taper which reduced the resistance considerably making it clear that eddy shedding was part of the problem. The preliminary report also discussed some tests with different surface finishes which will be discussed later.

The final report was dated 13th December 1872 and the table below, extracted from it gives some of the results for a varnished plank at 600 ft/minute. The power of speed is that at which resistance varies.

VARNISH SURFACE	Length of plank, or distance from cutwater, feet			
	2	8	20	50
Power of speed	2.0	1.85	1.85	1.83
Average resistance in pounds/sq ft	.41	.325	.278	.25
Resistance/sq ft of last foot	.39	.264	.240	.226

Except for the shortest planks, resistance varied at a power of about 1.85. The resistance per square foot fell as the length increased though the rate at which it was falling was quite small at 50 feet. These results applied, with only small differences, to planks coated with anti-fouling paints.

Froude also tried surfaces deliberately roughened with sand, made to stick by sprinkling it onto a warm paraffin surface. He tried fine, medium and coarse sand and the results were of different character from those of the varnished surface.

Chapter 13: Years of Achievement 1872–79

MEDIUM SAND	Length of surface, or distance from cutwater, feet			
	2	8	20	50
Power of speed	2.0	2.0	2.0	2.0
Average resistance in pounds/sq ft	.90	.625	.534	.488
Resistance/sq ft of last foot	.730	.488	.465	.456

The resistance varied as the square of the speed at all lengths and the resistance per square foot was, of course, much higher. Froude remarked: 'Looking at the subject in a speculative aspect, however, certain features of the results present perplexing anomalies.'

The tests with planks covered in tinfoil were perhaps even more perplexing with a low resistance per square foot but varying as the square of the speed to much greater lengths than the varnished planks, though at 50 feet there was little difference.

These anomalies can be explained today, at least in general terms. At the leading edge of all surfaces the flow is laminar, that is each layer of fluid moves smoothly over the next. This flow breaks down into turbulence after a distance which depends on the speed and surface finish and on an exceptionally smooth surface, such as tinfoil, laminar flow will persist over a greater length than on moderately rough surfaces. The exact nature of the law of friction for rough surfaces remains a matter of debate, though this is mainly over the numerical measurement of 'roughness'. There were a few other problems which were not apparent to Froude. He did not record water temperature which has a slight effect on friction. There is also an added drag at the edges of a plank which was not accounted for in his procedures.

The longest plank was only 50 feet long and was pulled at 800 ft/minute (about 8 knots) which may seem far removed from a 500ft ship moving at 20 knots or more. However Froude pointed out

> 'For it was at once seen that, at a length of 50 feet, the decrease, with increasing length, of the friction per square foot of every additional length is so small that it will make no very great difference in our estimate of the total resistance of a surface three hundred feet long, whether we consider such decrease to continue at the same rate throughout the last two hundred and fifty feet of the surface or to cease entirely after fifty feet; while it is perfectly certain that the truth must lie somewhere between these two assumptions.'

181

William Froude did not publish any further work on the friction of surfaces but he did re-examine his work and refine the way in which the results were used (ref 13.4). There are notes from 1873 at Haslar showing that some slight errors in speed measurement were detected and corrected for. In 1876 he produced two curves for use at Torquay to simplify the scaling of resistance from model to ship. These graphs showed:

(a) The resistance per square foot of a paraffin surface to a base of length at a constant speed of 400 ft/min in fresh water, to be used for models up to 22 feet in length.
(b) The resistance per square foot of a varnished surface to a base of constant length for a constant speed of 6 knots in fresh water and intended to apply to ship lengths between 30 to 500ft.

The power with which resistance varies with speed was taken as 1.87 for paraffin and 1.825 for varnished surfaces.

This was William Froude's last note on friction and there is no sign that he saw any need for further work on the subject. His 1876 curve differs little from the earlier, published curves at 400 ft/min, particularly for lengths of over 10 feet though for shorter lengths the 1876 curve gave slightly lower values of resistance. The power of speed with which resistance varied was changed considerably; the 1876 value was 1.87 for all lengths compared with 1.95 at 2 feet and 1.93 at 20 feet.

For varnished surfaces, the 1876 curves were identical with the published curves from 30 to 50 feet. It is not clear exactly how Froude extrapolated to greater lengths but in the passage quoted above he shows that the precise formula used hardly affects the results. Payne notes that the 1.825 power of speed is usually attributed to Edmund Froude as it was first published in his 1888 paper but internal records make it quite clear that William was using it in 1876. This does not preclude it being Eddie's idea though there is no evidence either way.

Prior to 1876 the normal method of estimating ship resistance was to calculate the frictional resistance of both model and ship at various speeds; the deduced residuary resistance would then be scaled to ship size by "The Law of Comparison" and the ship frictional resistance added to obtain the total resistance. This was a tedious process and "Skin Friction Correction" curves were created from the new curves. These curves gave the correction, in pounds for each 10 square foot of model surface, to a base of model length, for various ship lengths between 50 and 500 feet. Similar curves were produced for model speeds other than 400 ft/min. This method of dealing with friction was used until 1883 when it was modified to conform with Edmund's new, non dimensional plotting. (Chapter 14)

13.2.2 Wave Making Resistance

Froude explained how wave making was linked to resistance in

Chapter 13: Years of Achievement 1872–79

a paper to the British Association in 1875 (ref 13.5).

> 'Now waves represent energy, or work done; and therefore all the energy represented by the waves wasted from the system attending the ship, is so much work done by the propellers or tow ropes which are urging the ship. So much wave energy wasted per mile of travel, is so much work done per mile; and so much work done per mile is so much resistance, and this cause of resistance at least would operate with full effect even in a perfect (i.e. non-viscous) fluid.'

In 1877 Froude described some experiments to the INA (ref 13.6) with an exceptionally long model (fig 13.2) in which bow and stern were joined by a very long parallel section. This enabled him to see the separate wave patterns generated at the bow and stern almost in isolation. In describing this work, Froude explained some of the main features of wave making and how these could be used to advantage by designers.

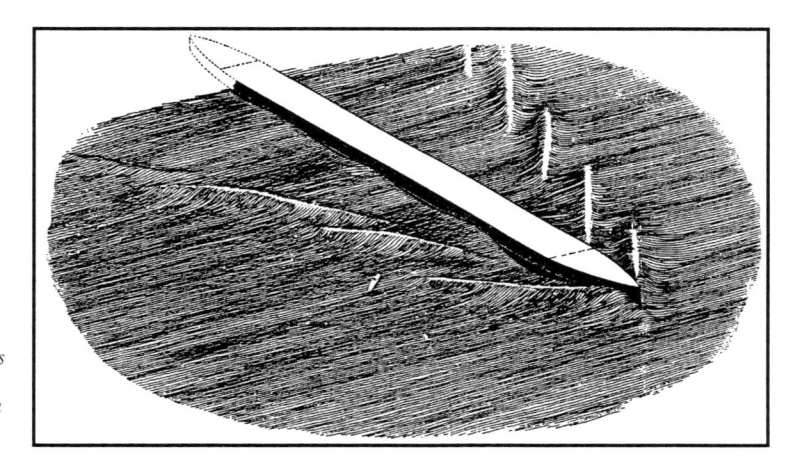

Figure 13.2 – Froude's sketch of the wave pattern generated by the bow of a ship. This model had a parallel section whose length was varied to study the interaction between bow and stern waves.

> 'Perhaps I had better say a few words more about the nature and character of these waves. The inevitably widening form of the ship at her "entrance" throws off on each side a local, oblique wave of great or less size according to the speed and obtuseness of the wedge, and these waves form themselves into a series of diverging crests, such as we are all familiar with.'

Froude then discusses the transverse wave system whose crests are at right angles to the fore and aft line of the ship, sketching the crests as they appeared against the side of his long model.

> 'It is seen that the wave is largest where its crest first appears at the bow, and it reappears again and again as we proceed sternwards along the straight side of the model, but with successively reduced dimensions at each reappearance.'

Froude then continues (summarised) – As the length of the wave crest from the model to its outer end increases, its height will diminish since the total energy in the wave, imparted by the

183

model, remains constant. The total wave making resistance is then the sum of the pressure forces which generate the diverging waves, the series of transverse waves originating at the bow and, finally, the stern wave system where the stream lines come together again, leading to an increase of pressure.

13.2.3 Eddy Making Resistance

Froude also introduced the term 'eddy making resistance' which allowed for the differences between friction over a flat plank and the total viscous resistance associated not only with the difference between friction over a surface curved in three dimensions but also included the loss of momentum associated with eddies behind the body.

> 'The mutual frictional resistance experienced by the particles of water moving past one another, combined with the almost imperceptible degree of viscosity which water possesses, somewhat hinders the necessary stream-line motions, alters their nice adjustment of pressures and velocities, and thus defeats the balance of streamline forces and induces resistance. This action is, however, imperceptible in forms of fairly easy shape. On the other hand, angular or fairly blunt features entail considerable resistance from this cause, because the stream-line distortions are in such cases abrupt, and degenerate into eddies, thus causing great differences of velocity between adjacent particles of water, and great consequent friction between them. "Dead water", in the wake of a ship with a full run, is an instance of this detrimental action.' (ref 13.7)

Froude's explanation of the cause of this component of resistance, today known as 'viscous pressure resistance', is clear and accurate, though he under-estimated its magnitude for fine forms. His choice of name misled many people over the years and it was generally believed that 'eddies' only occurred behind unfaired stern posts, shaft brackets etc or sudden changes of section such as bulges. All forms have some resistance from the cause outlined above and in those with bluff sterns such as wooden battleships and modern super-tankers this component can be very large and it does not scale easily from model to ship.

He considered one topic as needing further emphasis;

> 'I say distinctly, that the notion of head resistance, in any ordinary sense of the word, or the notion of any opposing force due to the inertia of the water on the area of the ship's way, a force acting on and measured by the area of the midship section is, from beginning to end an entire delusion.' – The area of the midship section was, and is, a convenient measure of the size of a ship – 'but it is an utter mistake to suppose that any part of a ship's resistance is a direct effect of the inertia of the water which has to be displaced from the ship's way..... The resistance of a ship, then, practically consists of three items – namely, surface friction, eddy resistance, and wave resistance'

13.3 Part II – General Hull Form

Froude was not a ship designer but he fully appreciated that design is a compromise and, that though hydrodynamics was important, there were many other topics to be considered.

> 'Do not, however, suppose that I shall venture on dictating to shipbuilders what sort of ship they ought to build: I have so little experience of the practical requirements of ship owners, that it would be presumptuous in me to do so; and I could not venture to condemn any feature in a ship as a mistake, when, for all I know, it may be justified by some practical object of which I am ignorant. For these reasons, if I imply that some element of form is better than some other, it will be with the simple object of illustrating the application of principles, by following which it should be possible to design a ship of given displacement to go at given speed, with minimum resistance, in smooth water – in fact to make the best performance in a measured mile trial.'
>
> 'The problem, then, to be solved in designing a ship of any given size, to go at a given speed with the least resistance, is to so form and proportion the ship that at the given speed the two main causes of resistance, namely surface-friction and wave resistance, when added together, may be a minimum.'
>
> 'In order to reduce wave resistance we should make the ship very long. On the other hand, to reduce the surface friction we should make her comparatively short, so, as to diminish the surface of wetted skin.'

Froude pointed out that this compromise had to be settled individually for each design and that there could be no question of carrying out the work on full scale but the series of model tests which he had started for the Admiralty were gradually accumulating the data needed for a sound judgement. He concluded the paper with some very strong statements of his philosophy and of the errors of the past.

> 'I wish in conclusion to insist again, with the greatest urgency, on the hopeless futility of any attempt to theorise on goodness of form in ships, except under the strong and entirely new light which the doctrine of streamlines throws upon it. ...... No one is more alive than myself to the plausibility of the unsound views against which I am contending; but it is for the very reason that they are so plausible that it is necessary to protest against them so earnestly; and I hope that in protesting thus, I shall not be regarded as dogmatic.'
>
> 'In truth, it is a protest of scepticism, not of dogmatism; for I do not profess to direct any one how to find his way straight to the form of least resistance. For the present we can but feel our way cautiously toward it by cautious trials, using only the improved ideas which the stream-line theory supplies, as safeguards against attributing this or that result to irrelevant or, rather, non-existing causes.'

It is interesting to compare these last passages with those in his correspondence with Newman on the need for evidence rather than dogmatic faith.

Froude's tests with SWAN and RAVEN had shown that there was no one ideal form and his new understanding of the different effects of friction and wave making on resistance helped to shown how the proportion and form of a ship could best be matched to its size and speed. Friction dominated low speed resistance where there was an advantage in short, fat forms with smaller wetted surface area. At high speeds wave making became increasingly important and increased length was essential. The proportions of a ship have a considerable influence on the weight of its structure and hence on its displacement so, as Froude showed comparison of a long, slender form with a shorter one is not simple; both have advantages and corresponding disadvantages and the choice depends on both the performance and on other requirements of the ship. Froude showed that if highly stressed by longitudinal bending in a seaway the structural weight (Ws) would vary as length cubed times beam (B) and divided by depth (D) (ref 13.8) or

$$Ws = (L^3 \times B)/D$$

In a real ship only a part of the structure is fully stressed due to longitudinal bending and a relationship of the form

$$Ws = L^x \times B \times D$$

is found appropriate. For modern frigates the value of the exponent of length, x, is about 1.3 but its value increases slightly with length though never approaching 3.

Reed's battleships were designed to carry a relatively small number of the new, bigger guns becoming available, in a short, 'centre battery' which was heavily armoured. In INFLEXIBLE armour thickness was 24 inches which, with 36 inches of teak backing, weighed 1100 pounds per square foot. In consequence, his ships were shorter in relation to their displacement than those designed earlier which reduced frictional resistance at all speeds but which could lead to excessive wave making at higher speeds. Froude was able to show that forms with a wide beam and a large midships section area linked with fine ends would minimise wave making at contemporary (low) speeds. To many, this was unexpected as traditional belief was that resistance was proportional to midships area.

In 1878, in a report signed by Edmund Froude, there is a comparison of the resistance of the INFLEXIBLE with that of AJAX brought to the same length and with a new form having a much greater beam but with the same length and displacement as the enlarged AJAX.

Chapter 13: Years of Achievement 1872–79

Name	Length	Beam	Drt	Dispt	Resistance in tons at		
	Ft	Ft	Ft	tons	12 knots	13 knots	14 knots
					0.67 $V/\sqrt{L}$	0.72 $V/\sqrt{L}$	0.78 $V/\sqrt{L}$
INFLEXIBLE	320	75	24	11,090	21.0	26.6	35.2
Enlarged AJAX	320	75.4	26	12,240	22.2	27.8	38.1
New Form	320	102.1	26	12,260	21.8	26.4	32.1

The wide form has not only less resistance at design speed but it also increases at a lower rate with speed. (ref 13.9) Such a form would also be less liable to capsize after severe flooding. Note that this result applies only to relatively low speed/length ratios, less than unity. (Speed/length ratio is defined as the square root of the speed V in knots divided by the length L in feet)

The Russian Admiral Popoff carried the approach to an extreme with two circular ironclads for coastal defence in the Black Sea intended to minimise both the extent of the armour and of the wetted surface. Froude ran a model of the NOVGOROD in the tank and showed that it could reach 8 knots without 'the formidable difficulties which she seems to meet at higher speeds' though even then her resistance was five times that of a conventional form. A 320 foot version could be driven to 14 knots provided it had sufficient power. Over a speed/length ratio of 0.8, violent pitching developed due to synchronism between the period of the ship and the wave. Though Froude remained fascinated by the limits to which very short, fat displacement ships could be driven, all his work showed that their limiting speed was very low. There were advantages, not least in safety, in some increase of beam for ships of moderate speed.

As already described, Froude tested a series of models with a long parallel body in which the waves generated at the bow could be separated from those generated at the stern. The basic model represented a fine form with a total length of entrance plus run of 160 feet. Other models had the entrance and run separated by a parallel body of lengths from 10 to 340 feet, giving a total equivalent ship length of up to 500 feet. Every 40 feet of length added 569 tons to the displacement which varied from 1245 to 5938 tons.

At low speeds the resistance was almost entirely due to friction and depended on the wetted surface area, virtually as length. At higher speeds the picture is much more complicated, the curve of resistance against speed being oscillatory with alternating humps and hollows. (Fig 13.2) Froude explained this by his observations of the positions on the hull side of wave crests and troughs generated at the bow. The maximum resistance always occurred with a trough 40 feet ahead of the stern while the minimum was when there was a crest at the same position.

Froude explained that the undulations in the resistance were due to variations in hydrostatic pressure over the after body or run. More simply, the crest of a bow wave will 'cancel' the trough which would normally be formed near the after end due to reduction of pressure under the middle portion of the hull whilst a bow trough will add to that formed aft causing a bigger resultant wave, draining energy from the ship. Later, Edmund gave the delightful description of the action of the stern trough in cancelling the bow wave as 'to swallow up that wave instead of making its own.'

William Froude also showed that at low speeds, the residuary resistance was independent of length and he suggested that this represented the work done in generating the divergent waves at bow and stern, so visible in his pen and ink sketch (fig 13.2) which did not interact with each other. It is interesting that these tests have recently been simulated in a computer with a similar picture of wave patterns.

In the 1870s it was quite common practice to lengthen cargo ships, increasing their capacity, and it had been noted that in many such cases there was an increase in speed whilst, in others, speed fell. This work of Froude explained which increments would be advantageous and also gave guidance on appropriate lengths for new ships.

The repeat tests of SWAN and RAVEN at Torquay showed the value of bluffer endings at higher speeds. This lesson was reinforced in tests with the torpedo vessel POLYPHEMUS which had a submerged torpedo tube in the forefoot whose outer cap formed a ram. The position and size of this ram had a marked effect on resistance as it formed what would now be seen as a bulbous bow. For more conventional forms, Froude recommended -"U" sections forward and -"V" sections aft.

In his evidence to the Commission of 1872, Froude did not think then (1872) that there was as much scope for improvement in form and hence speed as there was for reduction of rolling.

> 'I think it is not exactly the same, because I believe that no very great improvements are to be made in form. There are a great many questions requiring careful investigation. I have reason to believe that all the popular beliefs about the advantages of particular forms are mixed up with great errors and misconceptions of the fundamental principles on which the question depends; and mistaken beliefs on what really happens; but I think with certain limits the form of a ship is not so material as it has been considered. A particular form will do better at one speed, and another form at another; but still the

Chapter 13: Years of Achievement 1872–79

differences are not nearly so large as have been conceived, and certainly many beliefs on the subject are highly erroneous..........but I think that the general inclination amongst civil engineers who have not paid attention to the subject is rather to discountenance small scale trials, in the belief that they are extremely open to error; and so they are, unless the true principles of application are rightly considered.'

One may suspect that he would not have been so confident that there was little scope for improvement after the first few years of experiment.

13.3.1 *'Fall of ends'*

As a model or ship increases its speed through the water its trim will change and it will move bodily up or down; usually the movement is down until very high speeds are reached. Froude arranged recording drums on the resistance dynamometer on which movements of the bow and stern – 'fall of ends' as he called it – were recorded by pens driven by levers so that overall sinkage and trim (the change in draught between bow and stern) could be measured.

A grid was also painted on the side of the models so that the wave profile at various speeds could be noted. This profile was of value in deciding on the depth of the armour belt and in positioning fittings such as the hawse pipes, openings etc.

In developing the form for the 'Flat Iron' gunboats (known to Froude as the COMET class, usually now listed as the ANT class) testing began with a model of the Crimean War gunboat SNAKE, reduced to represent a ship 85 feet long. Twelve more forms were tested, all of 85 feet length and 256 tons displacement but mostly with 30ft beam instead of the 26ft of the reduced SNAKE. The design speed was 9 knots at which the original SNAKE form had a resistance of 1.98 tons, the first wide beam ship of 1.5 and the final version of 1.28 tons. There was also a requirement that these small gunboats should be capable of being towed at high speed by a bigger ship and Froude concluded:

> 'The "tip" of the ends of the very U-bowed designs (designated) DS2 and DV would be serious in amount and would obviously be fatal to any attempt to tow them at high speeds (ref 13.10). The tip of the ships of the flared bow design DU2 and the spoon bowed designs EB and EB2 would be considerably less than the others, and would be in the case of the two latter obviously (and in the case of the former probably) not such as to prevent their being towed even at 15 knots in smooth water.'

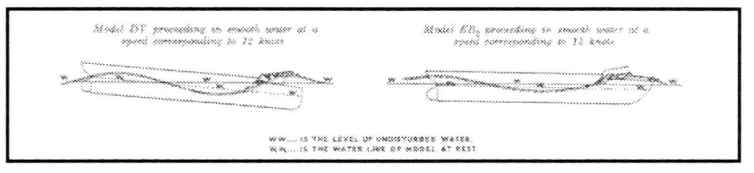

Figure 13.3 – Seaworthiness and low seaboard.

There are some intriguing, but not wholly clear, references in two letters of Henry Brunel in early October 1875 to a proposal, initiated by the Admiralty, to measure changes of trim on the full scale. Ships were to be steamed at high speed under the Saltash railway bridge which would form the base for observations. Two methods of measurement were considered; in the first a horizontal spar was to be suspended from the bridge which would strike, and mark, masts near the bow and stern. The alternative was to use a surveyor's level which was seen as difficult since the ship would be some 200 feet away. Henry suggested trying the techniques using a railway carriage under the bridge at Paignton but it seems that the proposal was quickly abandoned.

13.4   Part III – Propeller Performance –

13.4.1 Theory

Though Froude died before reaching a full understanding of the action of a propeller, he did make some important contributions which clarified some issues and, as described in the next section, he developed an empirical method for selecting the geometry of a propeller to suit a given ship and its machinery.

Froude's interest in propulsion problems was long standing and dates back at least to the model tests in Lake Bassenthwaite in 1850 (Chapter 3 and ref 13.11). Froude supported Rankine's approach adding remarks on the significance of 'wake', the way in which the movement of the hull drags water along with it so that the relative speed at the propeller is different from the forward speed and is irregular over the disk swept by the propeller. Behind a bluff ship the water is drawn along at virtually the speed of the ship giving a substantial pocket of 'dead water' which Froude illustrated by a the following story concerning an early wooden steam battleship.

> '...it was mentioned to me by the wardroom officer of such a ship that when the screw was hoisted out of the water, and the ship was under canvas going seven knots, he had descended into the well and bathed there without inconvenience.' (ref 13.12)

Based on his 1850 tests, Froude suggested that the propeller should be positioned some distance behind the hull if that was mechanically possible. He returned to this theme in a short paper which he read to the INA in 1867 (ref 13.13). Referring to the battleship story quoted above, he deduced that there was about 200 cubic feet of dead water weighing five or six tons.

> '...if we assume her to be driven at 10 knots by a two bladed screw, making sixty revolutions, the operation must be repeated twice per second, and, therefore, one hundred and twenty times per minute; the 200 cubic feet of water are dispersed with a

sternward velocity of about 10 knots, representing thus an extra thrust on the screw shaft of about 6000 pounds or 7000 pounds; while by the process of replacement, a retarding force of like amount becomes immediately impressed on the square tuck (transom) and stern post to which the dead water belongs.'

Froude did not mention it, but this extra force, 120 times a minute caused severe vibration leading to rapid wear in the stern gland. Leakage from the gland was so severe in the battleship ROYAL ALBERT that, on 29 December 1855, she had to be beached to prevent her sinking. (ref 13.14)

Froude joined in the discussion of a paper by Knowles to the Institution of Civil Engineers in 1871 (ref 13.15) in which he ended a long introduction praising the author's ability and learning by saying that all his conclusions were 'erroneous'. He began by showing the flaw in the generally held view that the force on a flat plate moving through the water obliquely was proportional to the square of the sine of the angle. He pointed out that if the plate moved full face into the water, it would set in motion the fluid ahead of it whilst if it moved obliquely, as it would if forming part of a propeller, some or all would be moving into undisturbed water. He continued:

'Thus, when a vessel was working to windward, immediately after she had tacked and before she had gathered headway, it was plainly visible, and it was known to every sailor, that her leeway was much more rapid than after she had begun to gather headway. The more rapid her headway became the slower became the lee drift; not merely relatively slower but absolutely slower.'

This passage is of interest on two counts; firstly because the action of plane surfaces at angles of attack relate directly to the dream which he shared with Henry Brunel of building a flying machine (Chapter 3) and also because it is one of the rare occasions in which his love of sailing is brought into a technical discussion.

Froude also draws an analogy with single oar sculling as further evidence and goes on to draw attention to errors in all previous experiments on the subject, including Beaufoy's. He then shows that the pressure over such a surface is not uniform but is much greater near the leading edge and, again, he brings his sailing experience to bear on the subject. He explains that many sail makers cut the jib so that it is 'baggy' at the forward end and hence –

'...when the vessel which carried such a sail was 'close hauled', that was to say, when the wind struck the sail obliquely from ahead, say at an angle of 45 degrees with the keel, the general wind pressure which the reaction which the rest of the sail produced swelled out the 'baggy' belt of canvas, not simply to leeward, but also so much forward that an observer viewing it in a direction at right angles to the vessel's course could see the convexity protruding itself ahead of the bolt rope, although, from

191

the direction of the wind current as a whole, that part of the sail, when thus protruded by the internal pressure, must experience externally also a considerable direct pressure on its convex or (so to call it) leeward side.'

Finally, he demolished Knowle's views on slip mainly using Newton's second law of motion – action and reaction are equal and opposite.

Froude returned to the discussion of slip and the performance of propellers in 1878, with a paper to the INA (ref 13.16). A propeller blade is similar in geometry to part of the thread of a wood screw but this analogy cannot be pressed too far. A screw driving into wood will advance a distance equal to its pitch for every turn but a propeller in water will 'slip', advancing much less than its geometric pitch per revolution. It was widely assumed that slip could be directly equated with loss of efficiency but Froude showed that this was only correct to a very limited extent and the scope for reducing such losses was small. He showed that friction on the rough blades of the day was much more important. Today, it is recognised that loss of momentum, rotational as well as in the line of motion, in the slipstream is as important as friction.

For a propeller generally similar to that of WARRIOR (1860) the breakdown of losses at the propeller is roughly:

	%
Momentum in line of motion	5
Rotational momentum	4
Viscous losses (including, but not exclusively, friction)	10
Effect of finite number of blades	1

leading to an overall efficiency of about 80%

Froude also set out the velocity diagram showing the relative motions of water particles over a rotating blade element which was also moving forward. This led to what is known as the 'blade element theory' of propeller action, often seen as a rival to, or even in conflict with, Rankine's 'axial momentum theory'. Both were incomplete and described the overall problem from a different viewpoint but they were completely consistent and were eventually brought together by Lerbs and his team during World War II in Germany and, afterwards, at AEW, Haslar.

13.4.2 Propeller Performance – Model Testing
Even though a complete theoretical treatment of propeller design was still out of reach, it was possible to obtain results which could be used in design from tests of models. The GREYHOUND trials had confirmed Froude's view that only a

Chapter 13: Years of Achievement 1872–79

small part of the power generated in the boiler was effective in overcoming the resistance of the hull and much of his effort in the last few years was devoted to exploring this aspect. He developed a dynamometer to measure the power actually delivered to the propeller, discussed later, but he was also aware that much of the power loss occurred in the interaction between hull and propeller.

He devised yet another dynamometer (fig 13.4) which was

Figure 13.4 – The propeller dynamometer first used in 1873 – it remained in use until 1938!

mounted on a separate carriage towed behind that carrying the hull model. The model propeller was driven from behind and the spacing of the carriages and the height of the model drive were adjusted so that the propeller lay in the correct position relative to the hull (ref 13.17).

The model propeller was mounted on the forward end of a shaft, driven by bevel gearing, and supported from a swinging frame on the carriage (ref 13.18). The frame was mounted on a parallel motion linkage, constraining it vertically and transversely, but permitting it to move in the direction of travel so that the thrust developed could be measured by the extension of a spring and recorded on a drum. The spindle driving the model propeller shaft was turned by belts (initially leather boot laces) from multi-groove pulleys connected to the carriage wheels so that the ratio of rotational to forward speed could be selected. The tension in these belts was measured to obtain the turning force applied to the propeller and this also was recorded on the drum. For twin screws, there were two frames with a common driving belt so that the sum of the thrusts and turning forces were recorded.

This dynamometer was a typical example of Froude design

The Way of a Ship in the Midst of the Sea

and construction, crude in places, indeed its resemblance to agricultural machinery of the day is notable, but it worked well and went on working. It was first used for tests with the propeller of ENCOUNTER in 1873 (fig 13.5) and remained in

Figure 13.5 – The first (two-bladed) propeller and the last tested at Torquay.

use until 1938, the last tests being on the propellers for the MANXMAN class fast minelayers of World War II. It was finally retired because it could only test models up to about 10 inch diameter and because maintenance was demanding to ensure all bearings turned freely.

Froude (ref 13.19) described a simple experiment showing the significance of interaction between hull and propeller. The propeller was first run without the hull model in place and its rotational speed adjusted until it produced no thrust. The model hull was then put in the correct position relative to the screw and the carriages run down the tank at the same forward speed. It was found that the propeller was exerting a considerable thrust which, in Froude's example, was equal to 20% of the model resistance and the resistance of the model had increased from three pounds without the propeller to three and a half with it running. This experiment was repeated at the same forward speed and increasing rotational speed until there came a time when the thrust developed by the propeller exceeded the augmented resistance of the model hull. Depending on the hull shape, this augmented resistance was

40–50% greater than that of the hull by itself.

Froude, with his usual ability to break down a complicated problem into soluble parts, considered the performance of a propeller first in isolation and then behind the hull. In 1883, Edmund Froude (ref 13.20) explained that it was perfectly possible to deduce the overall propulsive efficiency of the model from a single experiment but there were several objections to such a direct approach. The effect of viscosity on the flow round both the hull and the blades of the propeller on a small scale would cause the model performance to differ by an unknown amount from that of the full scale. Even if this problem could be overcome, it would still be very difficult, because of the number of variables – hull form, propeller geometry, positioning etc – to deduce general rules for propeller performance from such direct tests. Breaking the problem down enabled systematic experiments to be carried out on a limited number of variables.

This remains (1993) a contentious issue and methods used in different tanks vary considerably as all are only usable approximations to an unattainable truth.

13.4.3 Hull Efficiency Elements

The efficiency of the propeller in isolation is simply defined as the power put into the water (Thrust x forward speed) divided by the power supplied down the shaft (Torque x rotational speed) and is about 70% for a modern frigate. Behind a ship this relationship is more complicated since the forward speed is replaced by the speed of the flow relative to the propeller which is different since the ship will drag water along with it. The difference between the speed of the ship and that of the inflow into the propeller is called wake and is divided by ship speed to give wake fraction, w (ref 13.21).

The action of the propeller reduces the pressure and alters the flow ahead of it and hence the thrust required for a given speed is different from the resistance of the hull by itself at that speed. Froude described this difference as 'augment of resistance' (ref 13.22) later re-defined as 'thrust deduction' (t). He then added a third coefficient which he called Relative Rotative Efficiency (RRE) to take account of the changes in efficiency of the propeller in a non-uniform flow. These three coefficients, known as Hull Efficiency Elements, slightly modified, remain in use today but their physical interpretation, their dependence on form and the way in which they vary with scale remain a matter of controversy.

For WARRIOR, typical of fast ships of her day, w was about 0.34 and t was 0.26.

13.5 Part IV – Trials

William Denny read a paper to the British Association in 1875 on the conduct of measured mile trials (ref 13.23), proposing that such trials should be run at five speeds, covering as much of the range of power from the engine as possible. The trials had to be run in calm water to give an accurate comparison with model tests. By running over a range of speeds, accuracy was improved since a random error in any one would be obvious and could be eliminated. The range of speeds also made it easier to identify the individual components of loss of power. Froude welcomed this work and with Admiralty agreement, ran in the tank a model of the MERKARA, the ship whose trial results had been given by Denny. The comparison between model and ship results enabled Froude to deduce a better breakdown of the way in which the latent power in the steam was dissipated.

At the lowest speed of the trials, the resistance to be overcome was almost entirely frictional and hence it was possible to extend the curve of indicated horse power back to zero forward speed. At zero speed a considerable amount of power was still being absorbed which Froude saw as a measure of that needed to overcome initial friction in the engine, constant at all speeds. Froude tried a similar approach on other ships and, though there was some scatter, he deduced that this friction was about 1/7 of what he defined as SHP (ref 13.24).

He then produced a breakdown of power losses for a typical ship:

Froude's Breakdown of Power	
	Approx. Value
1. That required to pull the hull through the water (resistance times speed)	EHP
2. Augment of resistance due to the suction of the propeller on the hull ahead of it	0.40 EHP
3. Friction on the screw blades. (Based on GREYHOUND free-wheeling trial)	0.10 EHP
4. Constant friction due to dead weight and tightness	0.143 EHP
5. Friction due to working load, taken as equal to dead load	0.143 EHP
6. Air pump power (from Tredgold) as fraction of shaft power.	0.075 SHP

Chapter 13: Years of Achievement 1872–79

Figures 13.6 and 13.7 – Two views of the dynamometer or 'brake' intended to measure the power delivered to the propeller seen under test on HMS CONQUEST.

Equating these components gave
SHP = 2.347EHP

Froude then added 0.1SHP for slip making
ihp = 1.1SHP or EHP = 0.387ihp which he found to be typical of many single screw vessels of the period.

13.5.1 Ship Dynamometer
Froude sought to obtain further information on the way in which power was used and, in 1878, a dynamometer of his design was made to fit on the tail shaft of ships by Messrs

Easton & Anderson at a cost of £460-5-6. The dynamometer was required to absorb 2,000 horse power at 90 rpm and it was soon apparent that the heat generated would be too much for a simple friction brake.

Froude designed a water turbine (ref 13.25) which was tried successfully on HMS CONQUEST during March and April 1880 by Edmund after his father's death, when it was able to absorb 2,000hp at 70rpm and at greater rpm by partially closing the flaps. The CONQUEST was chosen as she had a lifting screw facilitating the installation of the dynamometer. Henry Brunel wrote (ref 13.26) that the Froude dynamometer for marine engines was a great success. It was tried 'on a huge scale (2000 HP) by Ed at Keyham this week and fully realised the rate of efficiency predicted.' A similar machine 5ft 3in diameter at 90 rpm would be capable of 4000 HP. 'Everything worked very well.' Henry noted that one of 14ft diameter would absorb the power of the whole Navy! Despite the apparent success of this trial the dynamometer was never used again and was finally broken up in 1922.

Smaller versions were used to calibrate apparatus at Torquay and later at Haslar. On William's death, the patent passed to Hurrell, his eldest son, and the 'Froude Brake' was manufactured and sold in considerable numbers by Heenan and Froude. Hurrell was also to be responsible for the structural design of the Blackpool Tower and the similar structures at New Brighton and Wembley, the latter never completed.

13.5.2 IRIS

The IRIS of 1875 is often seen as the first modern light cruiser and was a most innovative ship, the first all steel RN warship and probably then the fastest ship in the world. Her hull form was developed in a series of model tests at Torquay in January and February 1875 and she was propelled by two 3,500 ihp Maudslay compound engines driving two shafts. (ref 13.27).

Figure 13.8 – HMS IRIS.

Chapter 13: Years of Achievement 1872–79

The initial trials in December 1877 were most disappointing as she reached only 16.4 knots instead of the 17.5 confidently anticipated. It was thought that the area of the four blades of each propeller was too great and further trials in February 1878 for which two blades had been removed from each propeller confirmed this, at least at the reduced power which could be accepted on the remaining blades, limited by strength. In fact, the problem was that the original propellers were of too large a diameter and pitch. These trials were backed up by Froude's model tests, which included 6 bladed designs, in March and April 1878.

New four bladed propellers were made with less diameter and area and increased pitch and, in a trial in July 1878, IRIS reached 18.6 knots. A further trial with a new design of two bladed propellers gave a marginal increase of speed but at the expense of heavy vibration. It is interesting that no one seems to have recognised that the final speed, one knot faster than the estimate, was as big an error as the original failure to meet the design speed. However, Henry was to write (ref 13.28) 'Be sure that Barnaby (DNC) and Wright (E in C) appreciate what Froude has done for them and are grateful.'

Froude set out a trials programme which included running the engine at full speed (90 rpm) with the shafts disconnected when it was found that internal friction took 400 ihp whilst it was also shown that friction in the shaft bearings accounted for another 170 ihp. Froude then estimated the frictional drag of the propeller blades suggesting, not altogether convincingly, that the original design took 1120 ihp and the final four bladed 420 ihp. There is an intriguing but not entirely clear reference in Henry's letter (ref 13.29) on how to put canvas on the propeller without getting a knobbly edge to the blade and so that the leading edge does not cut the canvas. This was presumably intended to measure the effect of surface roughness on propeller performance. Recent experience with trying to apply coatings to propellers suggest that a canvas cover would be torn apart by the time the ship left harbour and, not surprisingly, no records of this trial can be found.

Froude also examined the resistance due to the drag of the exposed shaft tubes and brackets using a large scale section, 2ft 6in in length. To ensure parallel flow over the test section large plates were fitted, each about three chord lengths long and many times the cross section of the strut (fig 13.9). The top of the test section was about 6 inches below water and a third flat plate was arranged just below the surface in an attempt to prevent wave making drag. A hardwood strut connected the test piece to the carriage dynamometer (ref 13.30).

There were many problems and neither wave making nor the resistance of the strut were entirely eliminated but the results bear comparison with modern tests of the drag of streamline

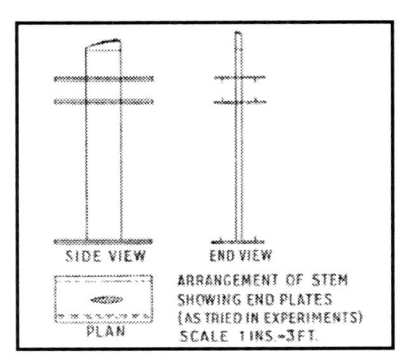

Figure 13.9 – Drag resistance.

struts. Due to scale effects, which can be large on such shapes, the resistance of full scale struts would be about half that of those which could be fitted to models, an approximation used by some tanks until quite recently. Froude devised some formulae, based on these tests, which were used at Haslar until long after World War II.

These tests were associated with others on a 1/24 scale model of IRIS, Froude showing that the resistance of the 'as fitted' elliptical section struts was about 10.5% of that of the hull, a figure which could be reduced to 6% by fitting a fairing at the trailing edge and with a sharp leading edge or 4.5% with a blunt leading edge – as Froude said 'It is blunt tails rather than blunt noses that cause eddies.'

The re-examination of the GREYHOUND trial included the effects of shallow water (Chapter 12) whilst trials in which she was allowed to drift before the wind gave an estimate of the magnitude of air resistance during trials. Later trials included measurement of speed through the air using an anemometer.

Wave profiles measured in the tank were compared with full scale during the trials of SHAH when Phillip Watts was hung over the side on a platform to record the ship's wave profile. Examination of the trials results of the SHAH showed a discrepancy and it was found that one of the two blades of her propeller had been set at 29ft pitch instead of the correct 24ft of the other. A somewhat similar problem was found in ENCOUNTER whose propeller distorted under load.

In a report dated March 1876, Froude examined the trials of the corvette OPAL and those of five of the ENCOUNTER class showing how errors could arise. He wrote that 'trials do not supply a very satisfactory test of the relative goodness of the form of different ships' but they are essential in throwing 'light upon other conditions of steam ships which model experiments taken alone cannot deal with.'

He showed that at 13.5 knots a straight run of 5,550 feet was needed before entering the mile to ensure a steady speed. If that distance was reduced by 1,000 feet the initial speed would be 1.5% and the average speed 0.2% low. Tidal variations could account for an eighth of a knot and if the errors were cumulative the result could be out by a quarter of a knot. Some of these errors were specific to Stokes Bay, where most warship trials were run and, a few years later, Edmund had the mile posts moved to a position where they could be seen more clearly.

13.5.3 Managerial Problems

Almost from the day that work started at the tank it was clear that the work was under funded and that there were too few professional engineers (see Chapter 11). Henry Brunel's letters reveal a number of other related problems. [Note – Henry is the only direct evidence for most of this section – only most important letters are referenced, others used] Writing to Izy in October 1873 (ref 13.31) he said that though the Admiralty were alive to the value of Froude's work – 'enough is done to make the Admiralty almost foolishly enthusiastic' – the experiments were not going as well as they ought and that if Froude wrote and asked for £1000 instead of £500–600 per annum they would agree.

Full use of the tank was not being made and experiments were interrupted, not for want of apparatus of workmen but because there was not the skilled brainpower available. He thought that there was work for '3 or 4 Eddy's' and noted that Eddy needed holidays. It was high time that Eddy's pay of £150 was raised and Froude should ask for more money and assistants. Henry thought that Hurrell would be very good as he has 'mathematical knowledge and would tackle the work better than Ed'.

This particular problem was exacerbated by several accidents and periods of illness which afflicted Edmund. In early 1875 he broke his arm near the wrist whilst roller skating and that same day there was an accident in the tank which caused serious damage to the steam engine. Later in 1875, Edmund suffered from a prolonged illness and Purvis had to be persuaded to stay on for an extra six months.

Some of Edmund's health problems were blamed on overheated and ill ventilated drawing offices in both the house and tank building. Henry also complained of the cold in the house as right through the winter all windows were kept open from breakfast to lunch and the fires let out and he describes 'the miserable and hopeless cold of the dining room at lunchtime'

William's health was not good either, Henry's letters refer to him being laid up by gout in 1869, 1873, 1874 and 1876 and there are indications of eye problems. Izy was very busy nursing Rose, Anthony Froude's daughter, a task so demanding that Izy became ill herself.

A related series of problems became apparent in 1875 when the Russians, Popoff and Goulaev wanted Froude to travel to the Black Sea to investigate the rolling of the circular iron clads mentioned earlier. In a very long letter to Froude, (ref 13.32) which can only be summarised, Henry outlined the difficulties, many of Froude's own creation. His starting point was that 'people should not work for nothing and especially professional men' though he admitted there were exceptions and he was prepared to see the founding of the tank as such an exception. Indeed, though Henry himself had the Brunel family's entrepreneurial sense, when working for Froude and the Admiralty, he asked for little more than his expenses. Froude's unpaid position at Torquay was right from his point of view but 'not right from the Admiralty view as they should not take goods without payment.'

Froude had placed himself in a position of responsibility to the Admiralty and the fact that they did not pay him ought not to affect his conduct and nor should the fact that they pay full expenses as 'Froude hasn't given them the chance.' He is a servant of the Admiralty in respect of his agreed work for them and for experimental and scientific work in which they place confidence in him and have placed themselves in his hands and many investigations depend wholly on him. At the same time Froude is a little in their debt as he has let them in for work which has cost much more and taken longer than expected.

This lack of any formal contract between the Admiralty and Froude implied a moral commitment on his behalf and in the specific case of the Russian work he should seek the formal permission of the Secretary of the Admiralty and not rely on Barnaby's verbal agreement. He should not work for Russia without payment – Henry suggested £1000 plus expenses – and he should remember that we had recently been at war with them and they remained our most likely enemies. He should also make it clear to Popoff that all results obtained would be passed to the British Admiralty. As an alternative, Henry suggested that he should himself go to Russia, suggesting a minimum fee of 500 guineas and £10-10-0 per day over 50.

As discussed in Chapter 11, funding was gradually increased, handicapped by Froude's lack of interest in money. In 1878 Barnaby, losing patience with lack of spending said 'Damn your eyes, for God's sake spend the money.'

13.5.4 Warship Design Assistance

The tank was built to develop better forms for RN ships and Froude recognised that ad hoc tests were an essential part of the programme while much of more general interest could be extracted from them. Some of the more important tests up to 1879 were:

GREYHOUND variants
ENCOUNTER & INFERNET (French)
FURY, INFLEXIBLE, DEVASTATION & DREADNOUGHT
TOURVILLE (French)
FANTOME
IRIS
TRAJANO (Brazil)
NOVGOROD (Russian)
OPAL
SHAH
COMUS
POLYPHEMUS
LIGHTENING
COMET
RAMUS Hydroplane

Chapter 13: Years of Achievement 1872–79

covered in some 60 formal reports to the Admiralty (ref 13.33)

The early tests with a model of GREYHOUND led into a systematic investigation of how the form of that ship could be improved. When the bow and stern were changed to the shape of the WARRIOR the resistance was increased by 8.5% at 8 knots and reduced by 16% at 12 knots with a small reduction in displacement (1161 to 1067 tons). Filling out the form amidships, increasing displacement to 1238 tons slightly increased resistance up to 12 knots but reduced it at higher speeds. Two yacht forms, based on SWAN and RAVEN, led to a top speed three quarters of a knot higher. Froude proposed in a letter of August 1871 to select a typical form and try the effect of enlarging the midship section for the same displacement and beam and then vary length and beam for the same displacement. Later he intended to alter the position of the maximum section and modify the lines at bow and stern (ref 13.34).

This plan of work shows a continuing difficulty of deciding which form parameters should be kept constant and which to vary. It is also clear that it would be quite out of the question to try such a range of modifications on the full scale.

Part of the programme took the form of a comparison of the British and French corvettes ENCOUNTER & INFERNET. The report dated 17th May 1873 is of particular interest as it shows the first ehp curves (fig 13.10) in which ehp, the power needed to tow the naked hull through the water, is plotted to a speed base. The considerable superiority of INFERNET was thought to be due to her greater length but in June further tests with a model based on INFERNET but reduced to the same length still showed the French ship as better. In fact, this was due to he fine ends associated with what is now called 'Prismatic Coefficient'. Henry wrote 'Model of French corvette which goes like hell compared with ours of the same class; nearly one knot

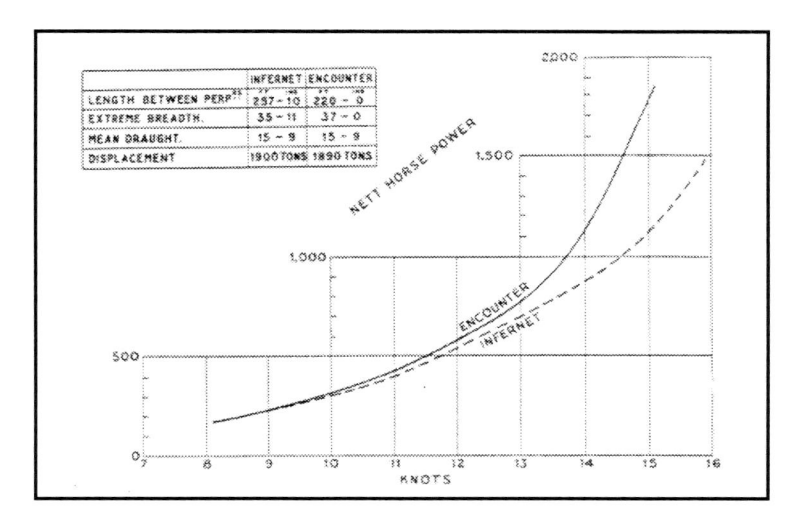

Figure 13.10 – First EHP curves from report dated 17.5.1873.

203

The Way of a Ship in the Midst of the Sea

faster – 15 to 14 knots.' (ref 13.35).

Prismatic Coefficient C = Volume/(A x L) where A is the area of the midship section

Froude was concerned that the fine ends might lead to pitching in a head sea but thought that a suitable disposition of weights would lead to acceptable motions. As part of the investigation of ENCOUNTER's performance, the first test of a model propeller, mentioned earlier was carried out on 15th August 1873.

The tests on FURY, later redesigned as DREADNOUGHT, in 1873 included some points of interest. Even though FURY was 320 feet long, her freeboard was only 4ft 6in forward in order to reduce the exposed area requiring armour protection. Froude tried a revised bow contour (fig 13.11) hoping that it would keep water off the fore deck but it was unsuccessful as, at 14 knots in calm water, the bow trimmed down by an amount equal to half the freeboard. When the freeboard was raised sufficiently to keep the deck dry, it was found that the ehp dropped by 12% at 14 knots. Froude also tried very large deflector plates to keep the bow wave off the deck (fig 13.12). These changed the trim at speed from 0.35 deg by the bow to 0.34 deg by the stern and reduced sinkage to about a third of that without the plates but they were impractically large.

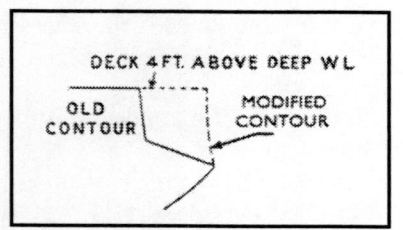

Figure 13.11 – Experiments on a model of HMS FURY, 1873 - Modified stern.

Testing a new hull forms soon became a matter of routine. A typical letter from Barnaby of 14th January 1879 reading (ref 13.36):

> 'Sir,
> I should be glad if you would favour me with your opinion of the horse power necessary to drive the New COMUS at 14 knots.
> May I ask for a reply at your earliest convenience.'

Since the model could be made in one working day and tested the next, the reply was usually very prompt.

13.6 Froude and the RAMUS Form

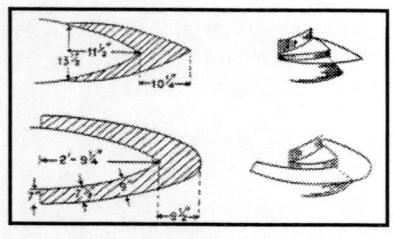

Figure 13.12 – Experiments on a model of HMS FURY, 1873 - Deflector plate.

One investigation which took up much time with, at first sight, little in the way of valuable results was on a stepped hull hydroplane, proposed by Reverend Ramus. This investigation is, however, worth studying at some length as it extended the range of experiments to very high speed giving further proof of the Law of Comparison whilst, at the same time, showed the faith of the Admiralty and Parliament in Froude's work (ref 13.37).

The Reverend C. M. Ramus wrote to the Admiralty on 8th April 1872 proposing that the speed of all steam vessels would be increased – 'at least double' – by adopting a form which he had devised based on a double wedge. Ramus met Barnaby two days later and on the 12th a letter went to Froude asking him to

Chapter 13: Years of Achievement 1872-79

include model tests of the Ramus form in his programme.

The following day Froude replied saying that he had intended to look at such forms anyway but querying the slope which Ramus proposed for the wedges, shown on his sketch as one in three, a value which Froude showed could not work. Ramus then changed to 8 deg and a month later to 2–3 deg suggesting at the same time that 1500 hp would drive an 1100 ton ship of his form at 30 knots in ordinary conditions and 50–60 knots in extraordinary occasions.

The first model was completed at the end of June and Froude suggested that he devote a week to trying it out and ensuring that the tank equipment worked properly at the highest speeds of 1240 ft/min and 'under the heavy strain which in spite of Ramus' anticipation it is, I fear necessary to count on at such speeds'

Ramus was invited to stay as Froude's guest at Chelston Cross to watch the experiments from 11th July and was surprised to find that the model was made to the shape shown in his own sketch of May as he had forgotten to tell anyone that he had later chosen a radically different bow shape. fig 13.13).

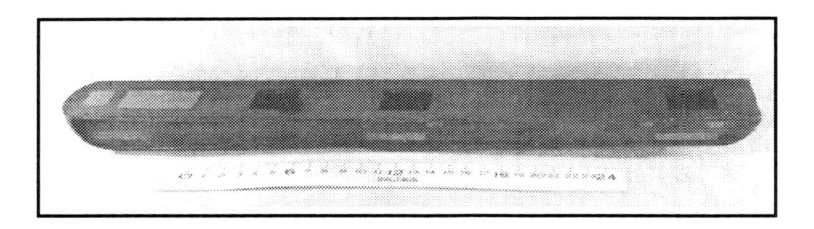

Figure 13.13 – Model of the RAMUS form.

Froude had two models of the RAMUS form

Scale	Length (ft)	Weight (lbs)
1/36	10.3	120
1/108	3.44	4.4

The larger model was originally of the shape proposed in May but altered later, with some benefit in resistance at lower speeds, while the smaller model was needed because, by the Law of Comparison, it would at the maximum speed of the carriage, correspond to a much higher ship speed than the larger model.

However, the results of tests with these models differing in size by a factor of three gave further confirmation of the Law for a form very different in character from those of SWAN and RAVEN and operating in a very different speed regime. After correcting for the effects of skin friction, the resistance per ton of the two models was almost identical at corresponding speeds. In addition the rise and fall of both ends of the two models were recorded during the experiments and sketches were made of the associated wave profiles which were also similar at corresponding speeds. Indeed, these tests are a far more convincing proof of Froude's Law than were those on SWAN and RAVEN, probably due to the much more favourable conditions in the tank compared with those in the river.

The resistance characteristics of the Ramus form at 2500 tons were compared with those of two other models, one which he called the PARA-RAMUS with the same dimensions and displacement but of conventional form with another called the GREAT GREYHOUND with roughly the same length as the Ramus but the lines and proportions of GREYHOUND. The table below, abstracted from ref 13.1, compared the resistance of the three forms at different speeds.

Form	Length (ft)	Dispt (tons)	Resistance 1,000 pounds at (knots)			
			10	17.8	52	130
Ramus	360	2,500	32	164	854	1,140
Para-Ramus	360	2,500	17	96.9	665	
Great Greyhound	360	9,280	36	180		

Froude's conclusions can be summarised as follows:

The RAMUS was only lifted significantly at speeds of over 30–40 knots though at lower speeds there was a considerable trim by the stern which could be mistaken for lift, a mistake still made sometimes today.

At speeds of 70–80 knots there was a very sensible amount of lift and at 130 knots this corresponded to about half the displacement.

At all ordinary speeds and even at exceptional speeds the resistance of the RAMUS will be extravagantly large compared with that of ordinary forms up to 50 knots and will greatly exceed the PARA-RAMUS. Even at 130 knots the resistance is greatly in excess of that at 50 knots and has not begun to reduce.

Froude then adds some figures showing the increasing effect of air resistance as speed increased. At 130 knots it accounted for 7lbs/ton out of a total of 182. He also drew attention to the dangers of such speeds at sea. If a RAMUS travelling at 60 knots met a wave with a slope of 10 deg it would fly some 100 ft before landing and if it had rotated in flight, which was likely, the destructive shock would be greater.

He was still attracted by the idea of dynamic lift from wedge shaped forms and developed an alternative with three wedges. This avoided the trim problem of the RAMUS craft when the rear wedge fell into the trough from the forward wedge at low speeds. He found that the forward wedge of a RAMUS could

Chapter 13: Years of Achievement 1872–79

come right out of the water, somewhat increasing the resistance; a phenomenon which he investigated in detail (ref 13.38).

Froude used his three wedge configuration to investigate wedge angles, developing a simple theory which suggested 1/17 was optimum (about 3deg). At this and the minimum resistance would be about 2/17 of the vehicle weight. Reducing the angle would increase the wetted surface and hence frictional resistance.

A copy of this report was sent to Ramus who expressed himself completely satisfied with the experiments but objected to the interpretation on the basis that Froude's law was unproven at higher speeds though, in fact the experiments provided just that proof. He also confused power and resistance.

On 5th August 1874 Froude wrote to the Admiralty asking that before publication of his report, they should ' ...alter all the passages and words which if published might appear personally offensive to Mr Ramus. When I wrote the report, it appeared to me difficult to express to Their Lordships any sense of the extraordinary want of knowledge of the subject with which he had undertaken to deal, and respecting which he displayed such a confident assumption of superior knowledge.....' (ref 13.40)

What makes these experiments so interesting is that the Admiralty accepted Froude's report without any shadow of doubt and, later, the correspondence with his report was printed and submitted to Parliament with general acceptance. Ramus was very unhappy and accused Froude of writing a report 'calculated to deceive'. Henry (ref 13.41) said Froude had been writing to this lunatic – a very crafty fellow and false... Froude's thunder was very good thunder but rather a blunder to write.' Froude was now fully accepted as the authority and later letters from Ramus were merely acknowledged by the Admiralty. Despite this, some contemporary and even some modern writers

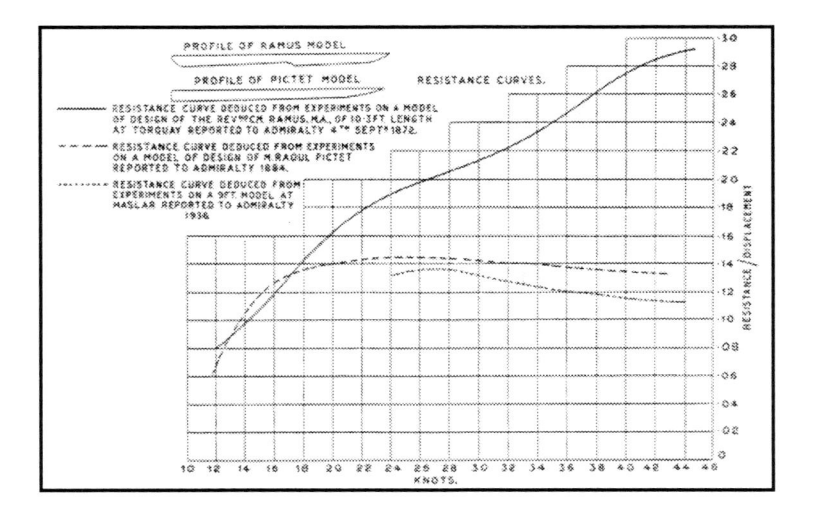

Figure 13.14 – RAMUS and other forms.

have portrayed Froude as reactionary; ignoring the advantages of the planing boat. This is nonsense; his own three float device was comparable to modern forms while the Ramus design was very much more resistful and quite rightly rejected.

Later, Ramus came back with ideas for a rocket propelled ram, claiming that a 50–100 ton craft of his design could reach 2–300 knots and that he had carried out his own tests to prove the claims. Froude, in a detailed reply showed that Ramus' tests were irrelevant – Reed quotes him as saying 'Irrelevancy could hardly be more apposite.'

Froude consulted Woolwich Arsenal on the performance of rockets and suggested that the Ramus ram might reach 12 knots for about 900 feet! Ramus also tried to get Froude's article for 'The Annual of the Naval School' suppressed as he was not given credit for his invention.

13.7   The Soaring of Birds and Things Aeronautical

Froude and his friends, Henry Brunel and Beauchamp Tower had long been interested in a flying machine. One aspect of this interest was Froude's study of the soaring of birds which was first apparent during his voyage to Greece in 1851. His last observations were made during his voyage to the Cape with Beauchamp Tower in 1879 and his thoughts were conveyed in the last of a long series of letters to Sir William Thompson dated 10th March 1879. The gist of this letter was published and, later, Edmund put together some more of his father's writing on soaring (ref 13.42).

He drew an analogy with slip in a propeller showing that a wing could only generate lift as a result of relative motion between it and the air leading to a differential pressure between top and bottom and hence, in truly still air, the wing would fall. Soaring of birds, with wings motionless, could only occur where there was an up current. He showed that many cases of soaring were seen on the lee side of the ship where turbulence could generate such up currents or at the head of an advancing weather front. Froude also showed that large waves could excite a resultant up current.

In one of his letters he made an attempt to quantify these views but it appears that the numerical values which he used were incorrect. The letters of early 1879 are of interest in a more personal sense since their combination of acute observation, careful analysis and light hearted style, suggesting that the depression which had afflicted him since Catherine's death the previous year had lifted.

13.8 Deaths

By the beginning of 1878 Catherine's health was giving serious concern; Henry Brunel writing to his mother 'Mrs Froude is in a poor condition and gets poorer and poorer' and he thought she would live only to the summer as the 'old are tougher than the younger ones'. Further letters (ref 13.42) describe further deterioration in her condition.

It was probably the approaching end of his wife's life that led William to put pressure on Edmund to marry and ensure the Froude's male line since, at that time, Hurrell was a widower with only a daughter. Henry wrote a long letter to Izy in June 1878 (ref 13.43) expressing the view that Edmund was old enough to make up his own mind and did it really matter if, in 30 years time, the Modbury estates were no longer owned by a Froude?

Catherine died of tuberculosis on 24th July 1878. William was greatly distressed and this, combined with the effects of long years of overwork without a holiday, caused a breakdown in his health. As early as March 1865, Henry had written of the Froudes '...they over exert themselves and suffer for it' and in October 1873 he wrote '...wish he take things easy a bit.' His friends pressed him to rest and Commodore Richards, in November 1878, invited him to visit South Africa in the corvette, HMS BOADICEA. Froude turned down this invitation at first but, under pressure from Henry, H Spedding and Richards, he eventually accepted. There was a suggestion that Henry should accompany him but he was too busy and, when Beauchamp Tower was taken ill it was arranged that he should go with William.

William Froude dined for the last time in London with his friends, Mary, Isambard (jnr) and Henry Brunel together with Edmund on the Saturday before sailing, going on to the theatre to see 'HMS PINAFORE' which Henry saw as appropriate. Henry wrote: 'glad he had a pleasant companion... in good spirits and looking forward to his holiday.

Henry, writing to Metford on 28th May 1879 tells the story of William Froude's last days. (summarised)

> 'Froude left in ill health ...He was certainly worn and fatigued by work and Mrs Froude's long illness but the trip originated from Captain Richards' invitation with the concurrence of all his friends. It seemed a good opportunity for a holiday as he hadn't had one for some years before and was not likely to have again, left in good health and spirits for an expected absence of about three months. The voyage was prolonged and was followed by some weeks quarantine, then he chose to prolong his stay at Capetown but throughout seems thoroughly to have enjoyed himself and his letters were most cheerful. He was about to return when illness seized him.'

He was clearly recovering as he wrote the note on soaring referred to above and had a final, lengthy tilt at Newman's views (Chapter 6). While in South Africa he contracted dysentery and died on 4th May 1879 at Simon's Town where he was buried with full military honours.

> 'In the absence of the Commodore, the arrangements for the funeral were made by Captain Adeane, who was assisted by Captain Wright and the other officials of the station. The coffin, a very handsome and very massive one, was drawn to the burial ground by a detachment of Marines, who were flanked on either side by officers. The Union Jack of England covered the coffin, and resting on it was a beautiful immortelle of chrysanthemums. The Government of the Colony was represented by Captain Mills, the Under Colonial Secretary, and there were several heads of departments present. All the vessels in port had their flags half-mast high. As the procession moved towards the burial ground, which is picturesquely situated on a slope, and looks down on Simon's Bay, the church bell tolled.' (The Times July 1879) (ref 13.43)

Initially, his tomb was marked by a simple iron cross, cast in the Dockyard but, in 1880, the Froude family erected a new gravestone for Catherine and her two children at Denbury and they had a similar stone made for William's grave. The inscription reads

> 'William Froude, Civil Engineer, FRS, LLD of Devonshire, England, died at Admiralty House 4 May 1879. In recognition of the great services which he had rendered to the Navy his remains were interred by officers and men in HM Ships in the port.'

and

> 'Lord, who shall die in thy tabernacle, even he that leadeth an uncorrupt life.'

Henry arranged transport across London and the stone was conveyed to South Africa in HMS TYNE.

There seems to have been a fear that those who had produced the iron cross might be upset when the family's cross appeared and Henry suggested the original cross should be left at the foot of the grave but it was decided to return it to England. Until 1960 it was displayed in the Superintendent's office of the Admiralty Experiment Works but it now lies in the old church at Dartington beside the memorials to his mother, brothers and sisters. Henry wrote an inscription which hung underneath which read:

> 'This cross was designed by Mrs Richards, late wife of Commodore Sir Frederick Richards, KCB, and was cast at her Majesty's Dockyard, Simon's Bay, Cape of Good Hope, and placed on the grave of Mr William Froude, who died at Admiralty House, Simons Bay, May 4th 1879.'

References

13.1 W. Froude. On some difficulties in the received view of fluid friction. British Association, London, 1869.

13.2 W. Froude. The fundamental principles which govern the behaviour of fluids, with special reference to the resistance of ships. British Association, London, 1875.

13.3 W. Froude. Experiments for the determination of the frictional resistance of water. Report to the Admiralty, London, 1874.

13.4 M. P. Payne. Historical note on the derivation of Froude's skin friction constants. Trans. INA. Vol. 78, London, 1936.

13.5 As 13.2.

13.6 W. Froude. Experiments upon the effect produced on the wave making resistance of ships by length of parallel middle body. Trans. INA. Vol. 18, London, 1877.

13.7 W. Froude. On useful displacement as limited by weight of structure and of propulsive power. Trans. INA. Vol. 15, London, 1874.

13.8 Report on the Committee on the INFLEXIBLE. HMSO, LONDON, 1878.

13.9 R. W. L. Gawn. Historical note on the investigations at the Admiralty Experiment Works, Torquay. Trans. INA. Vol. 83, London, 1941.

13.10 W. Froude. Supplementary note by Mr Froude to his remarks on Professor Rankine's paper. Trans. INA. Vol. 6, London, 1865.

13.11 W. Froude. Remarks on the mechanical principles of the action of propellers. Trans. INA. Vol. 6, London, 1865.

13.12. W. Froude. Apparent negative slip in screw propellers. Trans. INA. Vol. 8, London, 1867.

13.13 D. K. Brown. Before the Ironclad. Conway Maritime Press, London, 1990.

13.14 W. Froude. Remarks on: F. C. Knowles. The Archimedian screw propeller. Trans. ICE. Vol. 51, London, 1871.

13.15 W. Froude. On the elementary relation between pitch, slip, and propulsive efficiency. Trans. INA. Vol. 19, London, 1878.

13.16 W. Froude. Remarks on: A Holt. The progress of steam shipping. Trans. ICE. Vol. 57, London, 1877.

13.17 R. E. Froude. A description of a method of investigation of screw propeller efficiency. Trans. INA. Vol. 24, London, 1883.

13.18 As 13.16.

13.19 As 13.17.

13.20 Note: the wake fraction is defined in terms of the forward speed of the ship (V) and the mean speed of inflow into the propeller. (V+AΦ) Naval architects recognise two definitions; Froude wake fraction, w+FΦ, and Taylor wake fraction, w+TΦ. Though the Froude wake was suggested by Edmund, it seems that he normally used the definition attributed to Taylor. The Taylor definition is now universal. (1993)

$$w + F\Phi = -V - V + A\Phi\Omega, \quad w + T\Phi = -V - V + A\Phi\Omega V + A\Phi V$$

13.21 Note: there are also two definitions used to account for the difference between thrust (T) and resistance (R). R E Froude used 'Augment' (a) which was in use at Haslar until well after World War II when it was replaced by Thrust Deduction (t).

$$a = \frac{JT - Rj}{T}, \quad t = \frac{JT - Rj}{R}$$

In both Taylor wake fraction and thrust deduction, the denominator is the easier of the two quantities to measure. The relative rotative efficiency also contains all the errors involved in the process and is often seen as a factor of experience, correcting model results to ship values.

13.22 W. Denny. On the trials of screw steam-ships. British Association, 1875.

13.23 W. Froude. On the ratio of indicated to effective horse power as elucidated by Mr Denny's M. M. trials at varied speeds. Trans. INA., London, 1876.

13.24 W. Froude. On a new dynamometer for measuring the power delivered to the screws of large ships. Trans. Inst. Mechanical Eng. London, 1877.

13.25 The Brunel Collection, Bristol University Library. (28301 030480)

13.26 N. A. M. Rodger. The first light cruisers. Mariner's Mirror, 65/3, London, 1979.

13.27 As 13.25. (26317 100778)

13.28 As 13.25. (25343 160778)

13.29 As 13.9.

13.30 As 13.25. Note; the whole of this passage on Managerial Problems is based on Henry Brunel's letters, only the most important are referenced individually. There may also be a degree of bias involved in the single source but it seems consistent with what is recorded in the Froudes' Museum.

13.31 As 13.25. (19088 201075)

13.32 AEW Reports - 1872-1939, Ed D. K. Brown. Copies held in various libraries including National Maritime Museum, Science Museum and the PRO.

13.33 As 13.9.

13.34 As 13.25. (14036 010573)

13.35 Froudes' Museum.

13.36 Correspondence between the Admiralty and the Reverend C. M. Ramus on certain experiments conducted by them. Admiralty, London, 1873.

13.37 As 13.36.

13.38 As 13.25. (14105 150773)

13.39 Froudes' Museum.

13.40 As 13.25.

13.41 W. Froude. On the soaring of birds. Royal Society of Edinburgh, 1878.

13.42 As 13.25. Various.

13.43 The Times, London, July 1879 – quoting Capetown.

Chapter 13: Years of Achievement 1872–79

Tailpiece – The original tombstone for William Froude, made in Simon's Town dockyard. It was returned to England and hung for many years in the Superintendent's office before being placed in the old church at Dartington.

*William Froude, F.R.S., LL.D.
Pioneer of Ship-model Research, 1810–1879.*

The Way of a Ship in the Midst of the Sea

Chapter 14
Epilogue – Edmund Froude 1879–1919

14.1 Introduction

Figure 14.1 – Edmund Froude.

Even though Edmund had been in charge of the tank during his father's absence, the latter's unexpected death raised many problems, both professional and domestic. William's house, Chelston Cross was left to his eldest son, Hurrell, who decided to sell it. Everything had to be cleared and William's professional papers extracted from more personal material but this task was completed by April 1880. Edmund went into lodgings in Torquay at Belgrave Hall, (ref 14.1) which Henry Brunel did not find very comfortable. Some additional workshops were built on the tank site to replace those in the house.

Henry, himself, spent half his working hours for some months in writing to old friends of William gathering material for the obituaries (ref 14.2). He wrote obituaries for the Royal Society, INA and ICE and noted that they had to be different and was more than a little upset that the Royal Society published an incomplete draft. The Times published a letter from Dr. Acland and thought an obituary was unnecessary in consequence. It took some time for William's will to be proved and Henry had to lend the considerable sum of £1000 to Edmund. Henry envisaged a memorial volume containing all his professional papers including unpublished work and including an extended account of William's life, a project which reached fruition, at least in part, in 1955 with the publication by the INA of his collected papers (ref 14.3)

The Admiralty, understandably and quite properly, decided on a full review of the future of the tank before committing them selves to a successor for William. After some discussion they said in a letter of 5th December 1879 that proposals had been put to the Treasury covering the present year and also 1880–81, subject to termination by either party at 6 months notice. No work was to be undertaken which might extend beyond 31st March 1881 without special sanction from Their Lordships. The letter continued: (ref 14.4)

> 'Their Lordships do not wish it to be understood that they contemplate the severance of your connection with experimental work at the above date or the removal of the establishment from Chelston Cross at that date but being of the opinion that the conditions under which the experiments and observations of this kind should in future be conducted should now be carefully and deliberately considered, my Lords desire to reserve to themselves perfect freedom of action in order to make the best permanent arrangements for this service as soon as it may be possible to

arrive at a sound conclusion on the subject...'

Though the Admiralty's review clearly made sense, it was unsettling to Edmund, only partly offset by an increase in his pay to £12-12-0 per week. The total allocation was set at £2388 for 1879–80 and £2060 for 1880–81. All were conscious that, with William's death, they had lost their guiding light (ref 14.5). There was some relaxation, however, Henry, Edmund and others having a very enjoyable holiday in Henry's boat, Quicksilver, on the upper Thames in August 1879, repeated in 1882.

By then, in April 1880, Izy had married Baron Anatole von Hugel. Her husband was some ten years younger than Izy but had been well liked by William and their friends. His father was a distinguished Austrian diplomat and his mother Scottish. Anatole had made contributions to natural history and ethnology becoming, in 1883, curator of the Fitzwilliam Museum in Cambridge. It was a happy marriage and in retirement Edmund was to live with the Von Hugels. Edmund, and his work, suffered from a recurring ailment, said to be the result of a fall caused by tripping over a kerb in 1875 for which the treatment was opium and castor oil.

Edmund was to remain in charge of the Admiralty Experiment Works, as it became known in about 1890, (ref 14.6) for 40 years during which there were great changes both in warship design and in model testing. This long period during which William's work was extended and applied will be covered in this chapter, though in less space than that devoted to the seven years in which William ran the tank. Edmund's achievement will be discussed later but it should be appreciated that he advanced both science and its application and in no way was he a blind follower of his father. His theoretical work is not negligible though his real achievement lay in his contribution to the design of the ships of the navy in its greatest era.

14.2 Work Continues

Edmund and the tank remained at Torquay for seven years after William's death during which the main task was the routine but important tests of the new ships for the Royal Navy. Edmund devised a new system of plotting results, discussed later, and began the planning of his methodical series of hull forms. These activities led to some minor revision of the frictional resistance data which was re-plotted to make it easier to use with the new plotting. Edmund also carried out a few tests using models with roughened surfaces to investigate the effect on both frictional resistance and on the interaction (wake) with the propeller. In a paper to the INA in 1883 he thought that the application of such results might 'baffle us' and, indeed, there is still some doubt as to how to deal with rough surfaces.

As boundary layer theory developed it was realised that there was a flaw in the Froude data on friction as it was demonstrated that the resistance of a smooth, flat plate should depend only on the Reynold's number ($\frac{L.V}{v}$ Length x speed/viscosity) while the Froude data depended on length as well. The difference in practice is very small and the Admiralty Experiment Works did not change to the new system until 1967. For many years students at the Naval College were asked to show the inconsistency in the Froude method as an exam question. It is interesting that the small empirical correction factor needed to bring model data in line with ship trials (of the order of 0.97) appears to be a function of length; maybe Froude's powers of observation and deduction were in advance of theory (ref 14.7).

Only a few of the individual experiments can be mentioned. In 1884 a model of the WARRIOR, completed in 1861, was tested when consideration was being given to her modernisation. The model was tested both as a single screw form, like the ship, and with twin screws. In this case there were important differences between the propeller efficiency in isolation and the overall efficiency allowing for interaction with the hull. In isolation, the twin propellers had an advantage but, behind the hull, the large, single screw was superior. It is interesting that Rankine's formula, highly esteemed before Froude's work (Chapter 8), estimated the resistance of WARRIOR at 14.35 knots as 76,177 lbs while the model tests gave 49,000 lbs, an overestimate by Rankine of 55%.

In 1885 a series of passenger liner forms were tried, including the GREAT EASTERN, which was superior to all the other forms tried over the whole speed range, a credit to Scott Russell's skill and judgement. Edmund then designed a new form which was appreciably better. At the speed length ratio of liners, there was advantage in a fairly beamy ship but, in other tests on the new, fast torpedo boats, Edmund showed that great length was essential.

In 1885 some tests were carried out on a model of the new battleship EDINBURGH which had been found so hard to steer that, it was said, during exercises, she was told 'to take station on the horizon'. She was a very beamy ship, with a flat bottom and very full lines in order to minimise the proportion of her hull which required thick armour. Edmund deduced that eddies were being shed from the bluff stern, sometimes from one side, sometimes the other so causing her erratic behaviour. The ideal cure was to rebuild the stern, lengthening the ship 36 feet but this was thought too expensive and a compromise modification was adopted which seems to have been generally successful. It is now necessary to turn to some of the more general work of these last years at Torquay (ref 14.8).

14.3 Resistance, Hull Forms and Methodical Series Data

Edmund read a number of papers on resistance (listed in the Annex), some outlining the state of the art and others dealing with specific aspects such as frictional resistance and the way in which it varied with surface roughness. Tests of flat plates with different paint finishes were a small but continuing part of this work. In an early paper (ref 14.9) he extended William's work on the model with a parallel middle body, showing that there were critical speed/length ratios at which resistance increased very rapidly with speed which he called 'Humps' separated by 'Hollows' where resistance increased much less rapidly.

His most important contribution to the study of hull form and resistance was to develop a way of recording the form of models and their resistance characteristics which depended only on the shape of the model and which could be readily scaled to any size of model or ship. Graphs of resistance data in this form were, and remain, invaluable to the designer though today they are stored in the memory of a computer. The data from the routine tests of new forms were later supplemented by figures from a series of models in which the form was methodically varied.

He chose to define his unit of length as the cube root of the underwater volume which he called U. All other length dimensions were divided by U to give non-dimensional values while areas were divided by U^2. The shape of the model was then plotted on what was known as the elements of form drawing. It showed a curve of areas of transverse sections along the length, a midship section and the waterline shape, all in non-dimensional form and sufficient to give a reasonably precise definition of the shape.

Initially, resistance measurements were plotted in the form of a graph with resistance/ton on the vertical scale and a base of speed divided by volume to the power of 1/6. (equivalent to the square root of length used in William's 'corresponding' speed) The shape of such curves is inconvenient to use and Edmund eventually adopted a more complicated but generally more useful presentation in which the base was speed of the ship divided by the speed of a wave of length U/2, which he called circular K, and the resistance coefficient, circular C, was 1,000 times resistance per ton all divided by circular K^2 (ref 14.10). These 'circular' parameters can be truly non-dimensional but are usually written with numerical constants applicable only to Imperial units. To reduce the difficulty of handling all these peculiar functions, Edmund devised a special, double sided slide rule which remained in use until replaced by computers.

The graphs of circular C to base circular K were all drawn for a length of 400 feet. Since frictional resistance obeys a different law from that of wave making, a correction is needed for other

Chapter 14: Epilogue – Edmund Froude 1879–1919

lengths and a range of correction curves was included on the drawing.

The designer would use this material in the form of the Iso-circular K book in which each page displays resistance co-efficient circular C to a base of non-dimensional length for a single value of circular K. Each model was tested at three or four different displacements (draughts) giving a range of length co-efficients. Fig 14.2 shows how each page of the book contains a short line for each form, the totality of forms giving a broad trend of resistance with length/displacement ratio at that speed ratio circular K (fig 14.2).

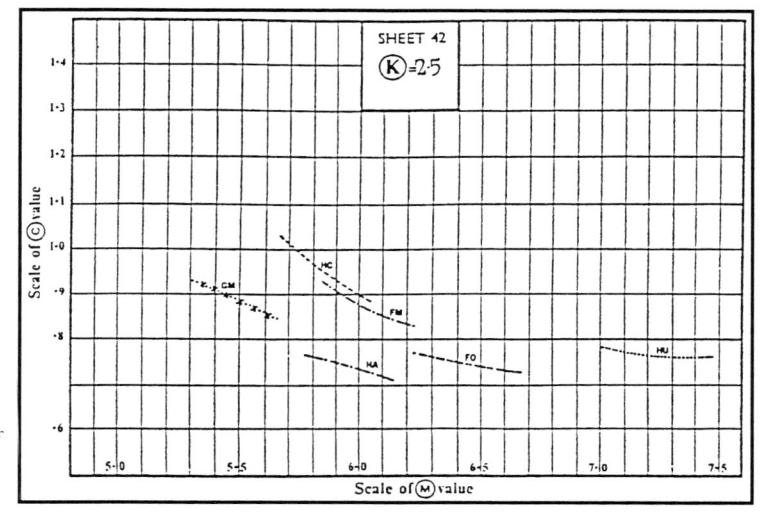

Figure 14.2 – One page from R.E. Froude's first Iso (K) book. (K) is a measure of speed, (C) of power and (M) the ratio of length to the cube root of immersed volume. Each line on the graph, eg FM, represents the results of one model run at several draughts and hence displacement.

At the intended design value of circular K the naval architect can select a curve which seems good, identify the model and hence turn up the elements of form diagram. This form may be suitable or it may have a distribution of areas along the length adapted to the requirements of the earlier design but inappropriate for the new ship in which case a new selection must be made. The Iso-(K) books now run into ten volumes covering tests up to 1973 and though largely superseded by a computerised data base, they retain some value for unusual forms.

Edmund published the first results of his methodical plotting in 1888 (ref 14.10). During the discussion of this paper Sir William White, the greatest of all Directors of Naval Construction, said 'It is not an uncommon thing for me to telegraph Mr. Froude that I am coming down to see him when some problem going beyond my experience has to be solved; and half an hour with Mr. Froude, in company with these diagrams, places me in a position of security which otherwise could not be

219

approached.' The value of these books was restated in 1940 when there was a possibility of a German invasion. a skeleton team from the design department was to be evacuated to Canada and the Iso-circular K books were first on the list of essential records to be taken with them.

Almost from the first, Edmund superimposed on the pages of the Iso-circular K book the results of a series of models whose shape was methodically varied from that of the torpedo cruiser VULCAN. He described six 'Types' of which the first used the basic form, the second and third had the after end reduced by 10 and 20 feet respectively whilst the last three had the fore body reduced by 5, 10 and 20 feet (from an overall length of 350 feet). The models were run at two draughts giving two beam/draught ratios. (57/22 & 66/19) The results, published in 1904 (ref 14.11) seem rarely to have been used directly but they made a convenient and consistent reference line against which the performance of specific forms could be compared. The idea was sound and led to the American, David Taylor, who had attended the Constructor's Course at the RN College, Greenwich, and had visited Froude, producing a much more comprehensive series which is used world wide today. (The USN ship tank outside Washington is known as the David Taylor Research and Development Center.)

14.4 The Move to Haslar

The tank at Chelston Cross had been designed for a two year life and even though it was to give useful service for several more years, it was apparent that the tank was too small and the building was deteriorating. The Admiralty letter quoted above shows that the possibility of moving was envisaged as early as 1879. The land owner, William Mallock, who had married William's sister, was well aware of the rise in land values in the outskirts of Torquay and 'much' increased the rent to £20 p.a. all making it clear that a new site was needed. The water supply, which had been free in acknowledgement of William's services to the company, would now have to be paid for.

By 1882, Edmund had sketched out the requirements for the new tank. It should be 400 feet in length with offices and workshops at one end (ref 14.12) He suggested that space be left for a 150 ft extension at one end which was finally completed 80 years later, in 1960. Many sites were considered including Charing Cross, in what is now Embankment Gardens, and a site alongside the Metropolitan Railway at Brompton Oratory where Cardinal Newman was now installed. There was also a proposal from Thornycroft for a jointly owned tank at Chiswick. Later, more detailed schemes were examined at Millbrook Penitentiary and in the extension to Portsmouth Dockyard but the final

Chapter 14: Epilogue – Edmund Froude 1879–1919

choice fell on the Gunboat Yard at Haslar which had been built by I. K. Brunel to outfit gunboats during the Crimean War (ref 14.13).

Figure 14.3 – The original offices at Haslar, seen after the Second World War, but little changed.

The move was carefully planned to minimise the disruption to the programme and to ensure that the new tank and its equipment would have a long and effective life. Edmund's plans were set out in a series of letters and memos during 1885. The last run at Torquay, number 46,190, was on Tuesday, 5th January 1886 and, to satisfy the book-keeping, the staff, who were on the books of Devonport Dockyard, were dismissed on Wednesday 3rd February, rejoining at Haslar on the Saturday. The sale in January of surplus equipment by a Torquay estate agent, Cox, raised £62-17-7, duly paid into Admiralty accounts. Edmund

Figure 14.4 – The original tank building at Haslar, now used as an accounting office.

Figure 14.5 – New tank at Haslar.

moved into lodgings with Mrs. Lindsey at 'Claremont'.

The detailed plans for the move covered the cost of travel, new engines, wax casting and cutting equipment and proposals for staffing the new tank. The ten staff from Torquay, all civil servants except Edmund, transferred and the new complement was:

2 Assistant Constructors (£160 pa)
6 Draughtsmen, shipwrights and apprentices (6/6, 5/- & 2/2 per day)
1 Writer (6/- per day)
1 Fitter (8/6 per day)
1 Joiner (6/- per day)
3 Labourers (3/6 – 4/6 per day)
—
14

The new tank at Haslar differed considerably from that at Torquay. It has a rectangular section, 20 feet wide and 9 feet deep and was 400 feet long as completed. Filling with water began on 15 May and took 12–15 days to complete. The wooden carriage (fig 14.5) under which the models were held, ran on rails at either side of the tank and was pulled by an endless cable. The carriage members were hollow box girders about 4 inches square formed of deal planks only 3/8 inch thick, screwed and glued together, designed with a wide overlap at the joints for a strong glued joint. This structure was remarkably strong and light, weighing only 6 cwt whilst being able to carry a load of 14 cwt, though given some steel reinforcement in later years. It remained in use until 1960 – as did the original fill of water.

The engine used to pull the carriage was a 10 inch Beauchamp Tower spherical steam engine (fig 14.6, ref 14.14) somewhat similar to the modern Wankel engine. Henry Brunel and Hurrell Froude were shareholders in the original company though it would seem that licences were granted to other firms. Edmund wrote to the Admiralty to say that he proposed to accept the tender from Heenan and Froude at £180 even though it was slightly dearer than the other tender as it was so much more convenient to deal with his brother! The spherical engine was often used to drive dynamos, mainly on railway engines though some were fitted to warships and Edmund thought that its characteristics would make it suitable for accurate speed control using a governor of his own design, based on that used at Torquay. The full power of the engine was needed only for acceleration to high speed. Most experiments required speeds in the region of 100–500 feet per minute but the carriage had been run up to 1200 ft/min (14 miles per hour). It is surprising how fast these quite modest speeds seem when standing on a girder

Figure 14.6 – A model of the rotor of a Beauchamp Tower's spherical steam engine; it turned inside a spherical casing.

bridge moving down an indoor railway.

Much other new equipment was obtained including a new wax model casting box and a cutting machine (ref 14.15). The cutting machine was based on the earlier one but was improved in detail. The cutters were moved up and down to new waterlines rather than moving the model which meant that greater support could be given and a number of features were incorporated to improve accuracy. The old spring steel templates were abandoned – there is a hint that this may have been whilst still at Torquay – and a pointer followed the waterlines on a long drawing.

The tank had to be drained for repairs to leaks during August 1886 but was soon refilled; the remainder of 1886 and the early months of 1887 were spent on calibration work, Henry, impatient as usual, commenting in October that it seems to take some time to get authority to do things.

14.5 Standard Models and Consistency

As early as 1880, Edmund had become aware that the measured resistance of a model appeared to change from week to week if not from day to day and, before starting new investigations, he wanted to be sure that the results at Haslar would be the same as those obtained at Torquay. While still at Torquay, in 1881, he had introduced measures to reduce the currents set up in the tank by the models and had started to use screw logs on the carriage to measure the residual current. Between 1881 and 1885 he studied the effect of temperature on the expansion of wax models and also on measured resistance using a model of the fast cruiser IRIS, mentioned in the previous chapter.

Edmund first made a sentimental gesture by christening the new tank with a flask of water drawn from that at Torquay. The first model was run on 22nd April 1887, and, by an amazing coincidence, its designation, in strict order, was HA. This was followed by HB a copy of the IRIS model, FM, which had been tested at Torquay. The Haslar results for the new model were 5.6% higher than those obtained at Torquay, a difference which could not be explained. Edmund wrote 'It seems that it will be very desirable to keep a model the year round, trying it at intervals. The IRIS model naturally suggests herself for this'. Model HB lasted until 1890 when its surface was found to have deteriorated and it was replaced by a replica but only after scarphing experiments. In 1895 a brass model was made and was called IRIS though it bore little resemblance to the shape of the original and this model remains in use today as the standard and any correction needed as a result is called the IRIS correction, a replica was made for the National Physical Laboratory when their 'Froude Tank' opened in 1910.

The Way of a Ship in the Midst of the Sea

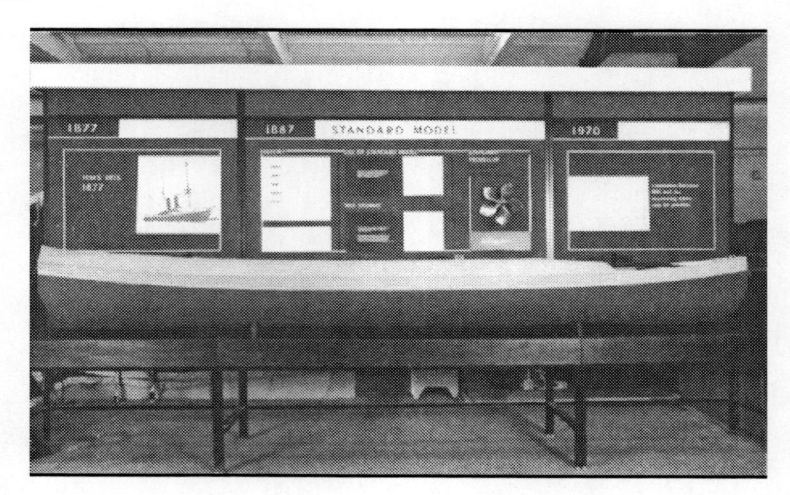

Figure 14.7 – The standard model IRIS which bears little resemblance to the earlier IRIS ship.

There are many causes for differences in resistance measurements from day to day and from place to place and occasionally there were violent changes known as 'IRIS Storms', though there is no direct proof, it now seems certain that these were due to the presence of very long organic molecules of biological origin. Even a few parts per million can dramatically change the resistance (ref 14.16). a letter of Henry's (ref 14.17) says that Ed is having extraordinary results with the water. 'It's now discoloured and changeful in its friction.'

Edmund started trawling the water with a wide and fine muslin net with a record catch weighing 11 pounds (ref 14.18). Later, and until 1960, the water was kept pure by eels living in the tank which seems to have been reasonably satisfactory as the water in the tank was approved as fit for human consumption when the tank was nominated as an emergency drinking water supply in both World Wars. In 1941 one of the eels was found dead and though the probable cause was old age, the possibility of starvation could not be ruled out since, in peace time, they had been fed by the staff with scraps from their sandwiches. Each eel was then given a quarter pound of minced horse meat per week. In the late 1930s an apprentice put a pair of goldfish into the tank which bred furiously and buckets full had to be removed. Since 1960 the tank water has been treated with chlorine and there have been no 'storms' since.

Edmund Froude had an obsession with consistency; he was to tell the young Goodall, later Sir Stanley Goodall, Director of Naval Construction, 'In engineering, uniformity of error may be more desirable than absolute accuracy'. Goodall's comment, in his retirement lecture (ref 14.19) is of almost equal interest 'That sounds a heresy, but think it over.' Consistent errors can be allowed for using a correction factor but results which may either be exact or may be in error are more difficult to handle.

The joiner, soon after Edmund's retirement, took a different

Chapter 14: Epilogue – Edmund Froude 1879–1919

view of this love of consistency. On being told that his latest work was not up to Mr. Froude's standards, he replied 'Yes Sir, he was an old fusser.' (ref 14.20) It is also recorded that Edmund's bicycle was brought into the workshop each month for a check on the friction in the moving parts and any necessary remedial action. The bicycle was equipped with a number of dangling bells to give warning of his approach and carried a specially large net carrier on the handlebars, both of Haslar manufacture. Even after retirement, he would come into the works each week to check his watch against the works clock, itself calibrated against the time ball on the signal tower in Portsmouth Dockyard.

Edmund's professional work will be outlined topic by topic but it should be remembered that, in most cases, each topic spans 40 years work. In passing, one may note that he was elected to the Athenaeum in 1888, as a result of some energetic lobbying by Henry, which may be seen as marking his acceptance as an 'Establishment' figure.

Figure 14.8 – The workshop at Haslar. A modern safety inspector would go crazy!

14.6 Propeller Design

Edmund published a number of papers on the theory of propeller action based on the change of momentum of water passing through a hypothetical 'actuator disk', extending his father's earlier work (Chapter 13). This work did not give direct guidance on the geometry of a specific propeller but it did show how overall efficiency related to diameter, thrust loading and rotation in the slip stream. It also showed that half the increase in velocity of the water due to the propeller took place ahead of the screw. The 'actuator disk' neglects a number of losses afflicting real propellers and hence its efficiency represents an unachievable ideal which is still useful in discounting the many claims by over enthusiastic designers of novel propellers.

Such claims were even more common in the nineteenth century and there are many contributions by Froude in discussions at the INA in which he demolishes such over-optimism. He was quite willing to give credit to real advances by others, strongly supporting early work by F. W. Lanchester on which the modern Vortex theory rests.

The Froudes' early work on testing of model propellers has been described in the previous chapter, together with the way in which the performance of the propeller itself was separated from the effects of interaction with the hull. In 1881 Edmund read a paper (ref 14.21) describing the dynamometer and giving some preliminary results of the effect of pitch ratio, diameter and rotational speed on performance, concentrating on the interaction between hull and propeller. He discussed the apparent change in resistance when a propeller was working behind the ship and defined the term 'augment of resistance' though he rather confusingly also used the term 'thrust deduction' for the same physical phenomenon. Wake, the change in speed of flow behind a ship, was defined in a later paper in 1886.

Further papers were published in 1886 and in 1908, the latter paper covering a wide range of propeller geometries, pitch and diameter, and speeds, rotational and forward, with the results presented in different styles to aid design or trials analysis. This data was later extended but remained in general use in the UK until the Second World War. Once again, a Froude pioneered the use of non-dimensional co-efficients but again, as with so many pioneers, his work was later superseded by a more convenient presentation.

The effect of the direction of rotation of the shafts of twin screw ships was dealt with in 1898, arising from a very practical problem concerning the arrangement of the machinery (ref 14.22). It was thought that inward turning propellers would lead to a more compact machinery layout and Edmund was able to show that there would be a slight benefit in propulsive efficiency. In the early 1970s the author thought he had obtained a new insight into the action of twin propellers working in opposition to turn a ship at rest but further research showed that Edmund had reached the same conclusion some 80 years previously.

The importance of trials results received continuing attention and, in 1889, Arnulph Mallock (Annex 1.1) designed a vibration recorder which measured vibration levels in two planes at right angles using the principles of a seismograph. It consists of a

Figure 14.9 – The Mallock vibration recorder.

heavy brass ball, suspended from a long spring, attached to which are very light frames made from split cane which move pens across a roll of paper driven by clockwork (fig 14.9). This instrument was first used during trials of the cruiser MEDEA in June 1889 but seems to have been used only very occasionally after that.

14.7 Rolling and Other Problems

Edmund continued the work on rolling which his father had begun, and, in 1896, gave the INA a paper on the non-uniform rolling of ships. Whilst correct, it was not of great value until computers were available to carry out the very difficult calculations involved. He and Phillip Watts, the first constructor to work at Torquay, received the commendation of the Board of Admiralty for their development of anti-rolling tanks fitted in EDINBURGH and a few other ships. It may well be thought that these congratulations were premature as the tanks were not very successful and soon removed, the idea being re-discovered after World War II. The tanks on INFLEXIBLE were quite effective at small roll angles but were too small to help much at larger angles where bilge keels proved more effective.

14.8 Shaping Tomorrow's Ships (ref 14.23)

Edmund's greatest achievement lay in the application of model testing to hull design for the Royal Navy's ships which grew in number rapidly following the Naval Defence Act of 1889. By the time of Edmund's retirement in 1919 some 500 forms had been tested representing

        33 classes of battleship
        46 classes of cruiser
        61 classes of destroyer
        14 classes of submarine

This involved some 46,000 individual carriage runs at Torquay and a further 194,000 at Haslar. For each class there would be resistance tests on a preliminary form suggested by the design team, usually based on Iso-circular K book data, followed by tests on other, improved forms suggested by Froude after studying the results of the first test. Then there would be tests on the interaction between propellers and the hull, others to determine the alignment of shaft brackets and bilge keels and sometimes rolling and steering experiments. The scope of these tests increased year by year. Fig 14.10 shows the raw data, analysis and reports for one quite simple design.

Around 1900 there was a good deal of ill informed controversy

The Way of a Ship in the Midst of the Sea

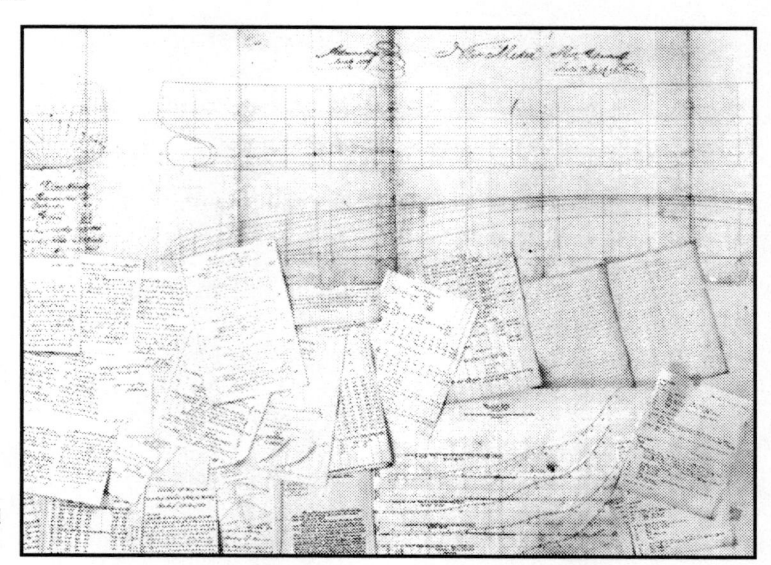

Figure 14.10 – A montage of the drawings, graphs and reports dealing with the tests for a typical ship; in this case the 'New MEDIA class', later the APOLLO class.

over the value of 'hollow' water lines at the bow. The fine ends seen necessary for low resistance at speeds such as those of small cruisers led to this feature but many seaman thought they also caused increased pitching. Edmund devised a wave maker consisting of a flapping board hinged at the bottom and driven by a crank off the workshop steam engine and used this in tests with models in waves, with and without hollow lines. The hollow lines were clearly superior in resistance in calm water and remained so in waves though their advantage was reduced. There was no measurable difference in pitching. Today one would choose a fairly full waterplane forward to minimise

Figure 14.11 – R. E. Froude (with hat) looking at the new wavemaker.

Chapter 14: Epilogue – Edmund Froude 1879–1919

pitching though, in most cases, the difference is not large. The wave maker remained in use until the late 20th century though, like grandfather's hammer, its components were replaced several times.

Though Edmund may have lacked some of his father's breadth of vision he could still draw conclusions from apparently routine tests which sometimes were far removed from hydrodynamics. For example, in 1881, he found that the beam of the cruiser NORTHAMPTON could be increased with only a very slight increase in resistance at operating speeds and, in a letter of 4th February to the DNC he suggested anti-torpedo bulges with a sketch almost identical to d'Eyncourt's very successful scheme of World War I (fig 14.12).

The roughness of various makes of anti-fouling paint was a

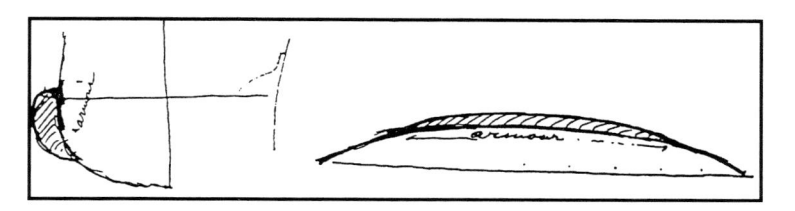

Figure 14.12 – R. E. Froude's idea for torpedo protection, very similar to the 'bulges' used during World War I.

frequent subject for study as was the effect of bilge keels. Work on torpedo hydrodynamics began in 1883 and was followed in World War I by work on paravanes to sweep moored mines and on hydrophones to detect submarines. Different forms of pitot and other types of 'patent' log were tried and there were many tests of the unusual external condensers used in steam launches.

There were tests of various patent propellers, usually without success. Froude tested models of foreign hull forms whenever information was available, which it usually was, since navies were then less security conscious than those of today. However, RN data from the tank for specific ships, as opposed to descriptions of procedures, was rarely published. He wrote a number of critiques of foreign tanks based on visits or published papers.

The first submarine model, of the a class, was tested in 1904 (fig 14.13) and by 1904 some elementary experiments on submarine control were being made. Free diving models were introduced in 1911. Regrettably, model tests had to be carried out in 1905 to explain the sinking of the submarine A8.

The first World War brought a big increase in work load shown by a comparison of the 1915 tests with those of 1906.

The Way of a Ship in the Midst of the Sea

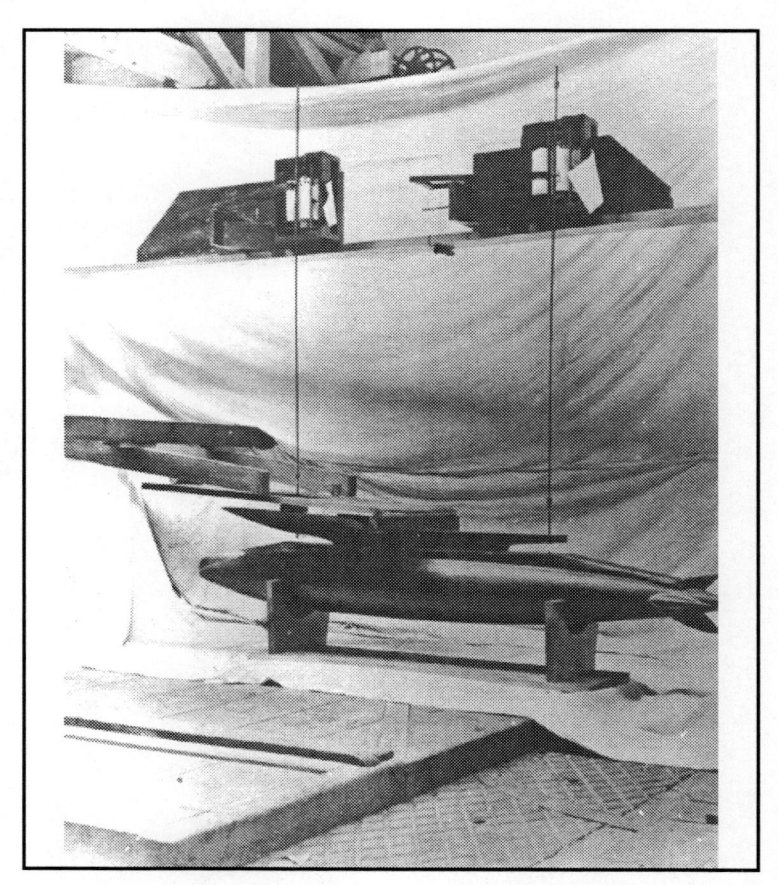

Figure 14.13 – The first submarine model test.

	Number of investigations reported	
	1906	1915
Surface ship resistance	12	50
Screw performance	4	11
Submarine work	4	11

In addition, 1915 saw tests on paravanes, anti-torpedo bulges and some novel forms. Since staff numbers remained at 14 the work load must have been formidable. Amongst the many novel tests during the war, were of the form of a destroyer with a 'straight line form' for ease of building. It proved little more resistful than its curvaceous sisters. There was also a series of tests to determine the increase of resistance due to the hole from a torpedo hit – surely increased resistance would have been the least of the captain's worries? When M. P. Payne took over from Edmund Froude in 1919, he found much of the apparatus obsolescent or worn out, mainly due to this heavy wartime load. One of his earliest purchases was a typewriter though it was to be another 15 years before a typist joined the staff.

14.9   Design Committees

Edmund Froude, like his father, was frequently consulted on broader issues of design and was a member of the 1905 DREADNOUGHT design committee. His advice on the performance of small, fast turning propellers when running astern finally tipped the balance in favour of turbine machinery. It was the weight saving from this novel machinery which made possible DREADNOUGHT's unique combination of high speed, heavy armament and extensive armour.

14.10   The Institution of Naval Architects etc

Edmund Froude read 16 papers to the Institution of Naval Architects, many of which have been mentioned earlier in this chapter. (Listed in Annex 14.1) He was a Member of Council from 1883 and Honorary Vice President from 1905. He took part in the discussion of other papers to a much greater extent than had his father and also contributed to other Institutions (ref 14.24).

14.11   Propagation of the Gospel

From the earliest days at Torquay, the Admiralty realised the value of a term of duty with Froude in the career development of its brightest naval architects and, starting with Phillip Watts, many young men spent two years at Torquay or Haslar. One of the greatest of all warship designers, Stanley Goodall, went to Haslar as a young man, learning several lessons. Like many others he saw Edmund as a 'great master of English.' (ref 14.25) On the other hand, Gawn (ref 14.26) a later Superintendent, says '... lacked the facility of simple and concise explanation that was so characteristic of the pioneer' (William). Many of R. E. Froude's reports are rather prolix and emphasised by Latin quotations. Even Goodall admits that trying to explain Froude's report on anti-rolling tanks to his senior he was told that it was a list of pros followed by a host of cons –

> 'Come back and tell me whether to go on fitting these tanks or not....Don't bring me a masterpiece of literature. Let your YEA be YEA and let your NAY be NAY.'

While Goodall was at Haslar, there was a famous legal action as some new destroyers had failed to meet their contract speed on one measured mile but exceeded it on a different mile. It was clear that the difference was the effect of shallow water and Goodall was sent to court to give evidence on the subject but the opposing Counsel had a point stronger in law. The contract said a measured mile and he had a letter from the Admiralty saying there was no objection to the (shallow) Maplin mile. Science was irrelevant (ref 14.27).

Commercial shipyards picked up the new technology in various ways. Some builders built their own tanks, (ref 14.28) others, like the most successful commercial warship builder, Armstrongs, relied on tempting Admiralty trained naval architects to work for them. William White and Phillip Watts were the best known but Perret, who followed Watts at Torquay brought with him a much longer experience of ship tank work.

Purvis, another of William Froude's early assistants became head of the 'Scientific Department' of William Denny's shipyard at Dumbarton in 1879. Two years later, he was in charge of the design and construction of a model tank in the yard, a near replica in size and equipment of the Torquay Tank. The first Superintendent was Mumford who had spent three years at Torquay and was to remain, with distinction, for 43 years at Dumbarton. Happily, this tank is now (1993) preserved by a consortium of museums. The front wall of the tank is of stone and carries a bust of William Froude. Writing about 1954, Gawn, then Superintendent at Haslar said (ref 14.29) 'I can confirm from personal visits that no Experiment Tank in Europe or America is complete without a plaque, photograph or bust in commemoration of the great pioneer.' The Garfield Thomas water tunnel in Pennsylvania started the fashion for christening a new facility with water from the original filling at Haslar which as mentioned earlier was christened with Torquay water. a very few privileged individuals, including the author, have been given a small phial of this magic fluid. Purvis later became a Professor in Japan and was instrumental in building their government tank in 1927. Mitsubishi set up an earlier tank in Nagasaki under two men trained by Denny.

The Spezia Tank was designed by Edmund and opened in 1898 under Rota. The St Petersburg tank, now the Krylov Institute, was also designed on advice from Edmund, opening about 1891. The Naval Construction and Armaments Company at Barrow (now VSEL) commissioned tests at Torquay, including the liner tests referred to earlier, and later at Haslar. In 1910 they, as Vickers, built a Haslar replica at St Albans. John Brown, after long debate, built a similar tank on Clydebank.

There were many other tanks and, in almost all cases, the designer came to Haslar for advice before starting, such as D. W. Taylor and the Washington Navy Yard tank. Finally,
after years of discussion, a British tank, open to all

Chapter 14: Epilogue – Edmund Froude 1879–1919

shipbuilders was set up at the National Physical Laboratory, paid for by Sir Alfred Yarrow. This tank, known as the Froude Tank was in use until recently and its first Superintendent, G. S. Baker, was yet another Haslar man.

It is generally true that all 150 ship tanks now operating, world wide, can be seen as deriving from the Froudes' pioneer work except perhaps, Dr. B. J. Tiedemann's Amsterdam tank (1874) which may be seen as an independent foundation (ref 14.30).

14.12 Yacht Racing Rules

Edmund was an enthusiastic yachtsman but, perhaps, gained more enjoyment from the framing of the technical rules governing the sport. Goodall (ref 14.31) says that the yachts which he designed showed to advantage when running free, but were not so successful when beating against the wind. In particular, Edmund played a major part in the restructuring of the racing formulae in use at that time in order to produce more seaworthy yachts.

Under the Royal Thames Yacht Club Rules of 1855 beam was heavily taxed but draught did not enter into the formula for tonnage and, naturally the beam of racing yachts reduced to the point where yachts had very little accommodation and poor seaworthiness. Such yachts were sometimes referred to as the 'leadmine' type from the weight of lead on their keel. Holt says (ref 14.32) that Froude attempted to attack the rule by reductio ad absurdum. The result was the JENNY WREN which had 7 tons of lead in the keel and only 5 tons register tonnage, a model of which was kept in the Froudes' Museum. Holt, a very enthusiastic yachtsman, says that she was very successful and would sail like a witch to windward but she had a remarkable angle of heel when close hauled in a breeze and was very, very wet. 'Some time ago I read a yarn by a man who had sailed in

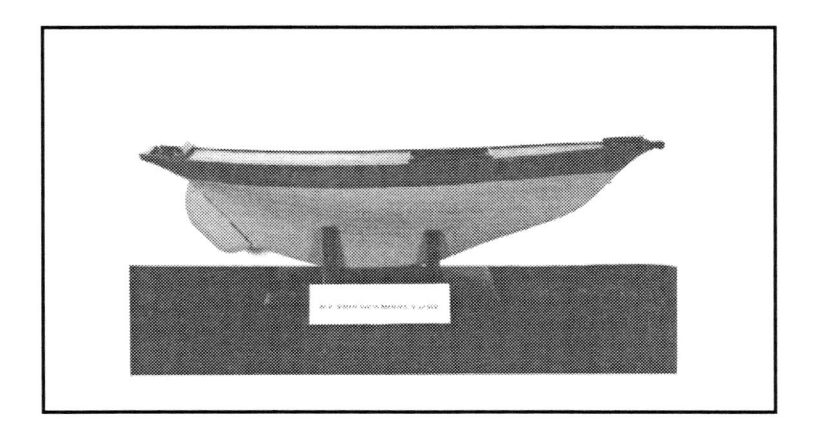

Figure 14.14 – JENNY WREN a yacht design by Edmund Froude.

this yacht. His description of hauling on the main sheet with the deck at an angle of 45 deg to the horizontal, a smother of water coming over the boat, and fear in his heart made very amusing reading.'

In 1886, after the loss on trials of OONA, an extreme example of this type of yacht, the Yacht Racing Association adopted a rule proposed by a British Naval Architect, Dixon Kemp, which depended on waterline length and sail area only. This very soon led to yachts with long overhangs, a 'skimming dish' form with high beam for stability and a fin keel to cut down wetted surface. Once again the racing rule had produced a yacht which was very uncomfortable in a seaway and due to lack of internal space was most unsuitable for cruising.

By this time it was becoming obvious that there was a need for a rating rule which governed not only the size of the boat but also ensured that it had a midship section which would give space for reasonable accommodation. In order to give advantage to the seakindly hull with soft bilges as compared to the skimming dish, Edmund, who had become a member of the council of the International Yacht Racing Union in 1896 (Later President) proposed that the skin girth be incorporated into the rating formula adopted by the yacht racing Union in January 1896.

$$\text{Linear rating} = (L + B + 0.75G + 0.5\sqrt{S.A})/2$$

In 1906 Edmund read a paper to the INA reporting on some 15 different rules which had been considered by an international conference. His major contribution was to highlight the problem, which is still with us today, that the successful racing yacht 'owes its success to what may be called measurement cheating qualities rather than general speed qualities.' Due note was taken by the conference that the rule finally chosen did try to take account of good seaworthiness and the habitability of the hull. This was achieved by the introduction of a new factor, the girth difference, (d), which was the difference between the skin girth and the chain girth, the latter being measured by stretching a chain from the deck at side round the bottom of the keel. The resulting rule was:

$$\text{Linear rating} = (L + 2d + 0.25G + \sqrt{\text{Sail Area}} - F)/2.5$$

This rule, thanks to Edmund's introduction of the girth factors, generally produced good, wholesome yachts suitable for both racing and cruising. Even today the rule under which 12 metres race includes a girth difference term.

14.13 Edmund Froude

Edmund built a house, North Lodge, which still stands as a hotel in Alverstoke, where he lived while Superintendent at Haslar. He was a shy and reserved person but well liked in the

Chapter 14: Epilogue – Edmund Froude 1879–1919

neighbourhood where he took some part in the local life, acting as manager of a local elementary school. However, he was the Master of Haslar and could be tough when opposed. During World War I, two of his assistants were called up for the army and a very strong letter went off from Froude saying they could not be spared and demanding that the calling up notices be withdrawn –it was. 'Who's Who' listed his hobbies as cycling, boating and photography.

His work was recognised by election to a Fellowship of the Royal Society in 1894, an Honorary Doctorate of Law from Glasgow in 1907 and the CB in 1911. By the time he retired his salary was £1,000 pa, roughly that of a Chief Constructor. He resigned in 1919 on medical grounds and the resignation was accepted by return of post, before a second letter withdrawing the resignation had reached the Admiralty. It was time to rest and time for new blood at Haslar. Edmund Froude was not entitled to a pension but it is believed that he was awarded a gratuity after representations by his successor, M. P. Payne. He continued to live in Alverstoke until 1923 when he moved to Cambridge to live with his sister, Izy, at Croft Cottage where he died on 19 March 1924.

14.14 Appreciation

Edmund's very real achievements will always be overshadowed by his father's genius. Some years ago, the author was discussing the work of the Froudes with two eminent naval architects and we speedily agreed that William was a genius whilst Edmund was 'merely' competent – and then we all laughed, wishing we were as competent.

When Edmund took charge of the tank, the largest ship in the Royal Navy was the INFLEXIBLE and when he retired, forty years on, the HOOD was just completing. a comparison of these two ships shows the magnitude of the changes in which he played such a large part.

Ship	INFLEXIBLE	HOOD
Displacement, tons, full load	11,800	45,200
Length, feet, overall	344	860
Speed, knots	14.75	31

His involvement was not confined to such prestigious ships but extended to every one of the vast fleets built before and during World War I, even down to the steam launches carried by the bigger ships. Submarines had introduced a third dimension well before he retired.

It is not easy to put figures to the benefit the Royal Navy

received from this work; in the introduction it was suggested that tank testing today reduced fuel consumption, on average, by some 5% compared with that of the preliminary form, with a very considerable cash saving. However, the true benefit is much greater as even that preliminary form relied to an increasing extent on tabulated data generated in the tank while the benefit in the early days must have been greater.

Fig 14.15 attempts to show how the resistance and hence the power per ton required to drive a battleship reduced over the years. The resistance coefficient measured during model tests of the final form is divided by that from Edmund's methodical series of models at the same length and corresponding speed. The downwards trend is clear and large even though some of the improvement is due to factors other than hydrodynamic. The development of improved material for armour reduced the pressure for unduly short ships whilst from the DREADNOUGHT onwards, the arrangement of more numerous guns led to increased length, both helping to reduce resistance.

The range of testing increased and even though resistance and propulsion work was still the main task when Edmund retired, there were more experiments on seakeeping, rolling, manoeuvring and on special tasks.

Edmund made some advances in theory, mainly in extending the momentum or actuator disk theory of propeller action but it is not unfair to say that he was more often up with the leaders rather than leading himself. Like his father, he was always concerned to pass on details of his methods to all, both to his staff and colleagues and, through the INA, to the world. Critics of his style of writing would suggest that his work might be more highly regarded had it been presented more clearly.

Edmund Froude was a true pioneer in the development and application of his father's work and the tributes from the Admiralty and other bodies on his retirement and, later, his death were well deserved.

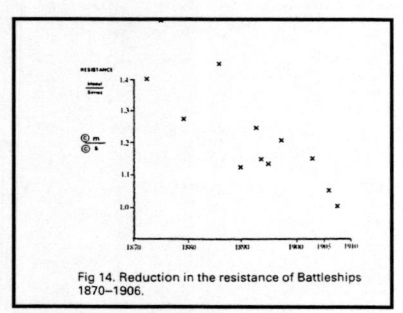

Figure 14.15 – The reduction in resistance of battleships from 1870 to 1910. Each cross shows the actual resistance of a model of a particular ship divided by that of the ideal from the methodical seris. There are, of course, many other factors which affect the form of a battleship.

References

14.1   Henry Brunel. 29013 120680 to Izy.

14.2   Henry Brunel letter books 27 to 29. As usual, only key letters will be referenced individually.

14.3   31092 291082 to Ed

14.4   Froudes' Museum.

14.5   27266 010679 'Since my father died nearly 20 years ago Mr Froude has not only the kindness of a father to me, but the confidence of a friend, I think I may say that I have had his opinion or advice in every step of importance I have taken'.

14.6   In 1890 the tank became formally known as the Admiralty Experiment Works though variations on this title had been in use for some time. Edmund Froude and his successors were very proud of the name and letters to the Admiralty Experimental Works were returned 'address not

Chapter 14: Epilogue – Edmund Froude 1879–1919

known' since the works was not itself experimental. Froude was given the title of Superintendent at the same time.

14.7 M. P. Payne. *Historical note on the derivation of Froude's Skin Friction Co-efficients*. Trans. INA., Vol. 78, London, 1936.

14.8 R. E. Froude. *HMS AJ,*' United Services Institute Proceedings, London, 1887.

14.9 R. E. Froude. *The Leading Phenomena of the Wave-making Resistance of Ships*. Trans. INA., London, 1881, Vol. 22.

14.10 R. E. Froude. *The "Constant" system of Notation of results of Experiments on Models used at the Admiralty Experiment Works*. Trans. INA., London, 1888, Vol 29.

14.11 R. E. Froude. *Some Results of Model Experiments*. Trans. INA., London, 1904, Vol. 46.

14.12 R. E. Froude. *Description of the Experimental Apparatus and Shaping Machine for Ship Models at the Admiralty Experiment Works, Haslar*. Trans. I Mech. E., London, 1893.

14.13 D. K. Brown, RCNC. *Before the Ironclad*. Conway Maritime Press, London, 1990. (See also Fig 14.5)

14.14 R. H. Heenan. *Description of the Tower Spherical Engine*. Trans. I Mech. E., London, 1885.

14.15 As 14.12.

14.16 R. N. Newton. *Standard Model Technique at the Admiralty Experiment Works*.

14.17 Brunel Collection 270687

14.18 R. W. L. Gawn. *Minore ad Majus*. A note prepared for the staff by the Superintendent ca 1954. (Froude Museum). A legendary history of the Admiralty Experiment Works - see Sources.

14.19 Sir Stanley V Goodall, RCNC. *Some RCNC Reminiscences* RCNC Journal 7. (Copy in NMM Library)

14.20 As 14.18.

14.21 R. E. Froude. *A description of a method of investigation of screw efficiency*. Trans. INA., Vol. 24, London, 1983.

14.22 R. E. Froude. *Experiments on the Direction of Rotation in Twin Screw Ships* Trans. INA., London, 1898, Vol. 40.

14.23 As part of the Centenary celebrations in 1972 a short film was made of the work of the Admiralty Experiment Works. Its title 'Shaping Tomorrow's Ships' was chosen as the best description of the work of the establishment.

14.24 K. C. Barnaby. *1860–1960 The Institution of Naval Architects*. INA., London, 1960.

14.25 As 14.19.

14.26 As 14.18.

14.27 As 14.19.

14.28 D. I. Moor. *Ship Model Experiment Tanks, the First Century - a British View*. The Parsons Memorial Lecture, NECI, Newcastle, 1984.

14.29 As 14.18.

14.30 J. M. Dirzwager. *Contribution of Dr Tideman to the Development of Modern Shipbuilding*. National Maritime Museum, Greenwich, 1981.

14.31 As 14.19.

14.32 R. W. L. Gawn. *Historical note on the Investigations at the Admiralty Experiment Works, Torquay*. Trans. INA., London, 1941. Discussion by W. J. Holt.

Annex 14.1 – Papers to the Institution of Naval Architects (by R.E. (Edmund) Froude)

Date

1881 The leading phenomena of the wave making resistance of ships.

1883 A description of a method of investigation of screw propeller efficiency.

1886 The determination of the most suitable dimensions for screw propellers.

1888 The 'Constant' system of notation on results of experiments on models used at the Admiralty Experiment Works.

1889 The part played in propulsion by differences in fluid pressure.

1889 Remarks on Professor Greenhill's theory of the screw propeller.

1892 The theoretical effect of the race rotation on screw propeller efficiency.

1892 Convenient curves for determining the most suitable dimensions for screw propellers.

1892 Some additional features in the 'Constant' notation used at the Admiralty Experiment Works.

1896 The non-uniform rolling of ships.

1898 Experiments on the direction of rotation in twin screws.

1904 Some results of model experiments.

1905 Model experiments on hollow versus straight lines.

1906 Yacht Racing Measurement Rules, and the International Conference.

1908 Results of further model screw propeller experiments.

1911 The acceleration in front of a propeller.

Other Papers

1887 Royal United Services Institution. HMS AJAX. The alterations lately made in her, illustrating the truthfulness of the results obtained by experiments on her model in the tank at Torquay,

1893 Institution of Mechanical Engineers. Experimental apparatus and shaping machine for ship models at the Admiralty Experiment Works, Haslar.

1894 Greenock Philosophical Society. Ship resistance.

Chapter 15
William Froude – An Evaluation

15.1   Achievements

William Froude is commemorated today in Froude's Law of Comparison and the Froude Number on which the law depends. His technique of model testing, with results reliably scaled to ship size, meant that, for the first time, it was possible to make accurate estimates of the power needed to drive a ship. Since it was quick and cheap to make and test a model, it became possible to develop for each ship a form which was suitable for its role, speed and size. Before Froude's work, power estimates were based on experience and while such estimates were usually reasonable for forms and engines close to a previous ship, those for novel forms were greatly in error, leading to increased cost and poor performance.

The first step was to estimate hull resistance for which Froude's law of comparison was first demonstrated in the SWAN & RAVEN tests and confirmed by the GREYHOUND trial. The SWAN & RAVEN also showed that there was no single, ideal form applying to all ships at all speeds as was then believed. Froude then refined his method by separating frictional resistance and providing very reliable data for its contribution to the total.

Propeller performance is far more complicated than resistance because of the number of variables involved. An essential first step in propeller design is to know the resistance of the hull which gives a first estimate of the thrust required from the propeller. Froude showed as early as 1850 that this value must be modified very considerably to account for the effects of interaction between the flow round the hull and that induced by the propeller itself. Froude designed a dynamometer which would measure the performance of a model propeller both by itself and behind a ship and showed how to scale the results. Once again, Froude's work became available just in time as the new compound engines, working at higher steam pressure and higher rotational speed, were less tolerant to a mismatch with propeller characteristics than were earlier engines.

These two developments, of resistance modelling and of propeller performance, meant that, for the first time, a ship could be designed in the confidence that it would reach its design speed and not be burdened by unnecessarily heavy and costly engines. Since William Froude and his son Edmund published all their basic work, model test tanks were gradually established by all leading shipbuilding countries, often with direct help in the design from the Froudes, and, whilst there are still arguments over the details of technique, they all depend on

Froude's Law of Comparison.

Froude's personal contribution to the understanding of resistance can be regarded as applied research based on observation and analysis of model tests. His papers make it clear that this work was soundly based on the most advanced theories of the day relating to the behaviour of waves and on frictional boundary layers.

Froude's earlier work on rolling differed in approach as he first set out a novel and generally comprehensive theory. He then developed a process of model testing and graphical analysis to give practical answers in aspects where the theory was incomplete or where the equations could not be solved. The theory was not easy to understand and the testing process fairly lengthy so that, as Froude himself said, rolling was not given as much consideration as it might have had, due largely to lack of appreciation of the loss of capability caused by motions and their effect on crew performance.

Full vindication of Froude's approach to rolling came only in World War II when the US built Captain class frigates of the Royal Navy were found to roll excessively. Using only Froude's methods, a cure was found involving bigger bilge keels, reduced metacentric height and increased moment of inertia. However, the cure was specific to this one class and a more general approach had to await the developments in probability theory and in computers after the war.

Froude's work led to a more general appreciation of the value of model testing which has since been much extended, both to other aspects of ship behaviour and to different dynamic problems. It should be remembered that, in 1870, the whole technical and scientific world was convinced that model testing had been fully explored and that it had been shown that such tests did not give valid answers. The members of the British Association 'Existing Knowledge' Committee were not reactionaries but were all very progressive engineers or scientists and yet none of them supported Froude's views on the use of models. Similar, negative views were held in other countries and only the Dutch engineer, Dr. B. J. Tidemann, was working along similar lines.

There was also an impact on thinking within the Admiralty, far more progressive in technical matters than is usually thought. They now had a research laboratory and successive Directors of Naval Construction were in the habit of posing novel questions with fair confidence that the Froudes would produce an answer. The author was able to respond to one such query, demanding model tests, circa 1960, by saying that the tests were not needed since unpublished work by William Froude provided the answer.

Both Froudes were excellent teachers and William's early assistants, Watts, Perret and Purvis, all rose to the top of their profession whilst other future leaders acknowledged the value of their early experience under Edmund. Naval architects were led into asking the right questions at an early stage of the design.

Today, no one would contemplate designing a ship without model tests to ensure that the form is 'fit for purpose' being economical in fuel consumption and a good sea boat while the propeller should be efficient and free from unseemly noise or vibration. It is no wonder that his portrait graces most hydrodynamic laboratories. Some contemporary tributes conclude this chapter.

15.2   Method of Working

The long correspondence with Newman provides many of the best indications to the way in which Froude's mind worked – or at least the way in which he thought it worked, which may not be quite the same thing. The frequent use of the phrases, 'a sacred duty to doubt' and 'probability is the guide of life' suggest both his willingness to overturn accepted theories and practices, shown in his early railway work such as his novel skew bridges and his paper on expansion of steam. There is also a hint of the self doubt which he occasionally showed as in his support for an advisory committee to monitor work.

A note by a life long friend, Frederick Rogers, (Lord Blatchford) is on a similar theme. 'William used to say that he had never mastered a principle until he had thrown it into a paradox.... And paradox, of course, invites contradiction and so controversy...'

Newman frequently suggested that scientific discoveries come as a sudden flash of illumination as in an act of faith but Froude always maintained that discovery took the form of a steady and methodical process of putting one fact on another perhaps best explained in the following extracts from his letter of 29th June 1853,

> 'the skein of nature is not to be unravelled by pulling at the loops which seem to offer themselves to be pulled at but by cautiously following the clues...' and

> 'generally speaking the great scientific discoveries are but the crowning results of some series of long continued and patient enquiries of men known as other enquirers ...'

This step by step approach seems at variance with his own description of his discovery of the importance of wave making in resistance to roll which supports Newman's argument. In his paper of October 1872 (ref 15.1) he wrote 'On travelling again and again over the whole question the idea suddenly suggested itself that the waves created by the oscillation had been left out of account...' His recognition of the significance of the movement of his small float in waves also seems more sudden than step by step. Froude would probably maintain that his sudden conclusion was merely associated with the last step of a patient enquiry.

He was a perfectionist, at least where perfection was necessary, with a real hatred of sloppy work and sloppy thinking. His description of a machine displayed at Falmouth in 1869 includes the words 'professional immorality' (ref 15.2) which is matched by James Spedding in his letter to Childers in April 1869 '...he is moved by scientific curiosity and the love of a perfect machine (especially a ship) and the sense of the immorality of wasted force and unnecessary friction.' (ref 15.3) One may see his use of immorality as a 'strong' word, reflecting his real hatred of sloppy work or thinking.

Froude had a great breadth of understanding, shown first in his wide ranging developments in railway engineering. Later, he was to show, quite frequently, an ability to draw conclusions from apparently unrelated facts. This breadth of understanding also meant that he could break down a complicated problem into components which could be solved individually whilst still keeping the overall solution in mind. In the following section, Froude's method of working is considered under various headings and, while this breakdown is convenient, it should be remembered that Froude would often use more than one approach simultaneously.

15.3 Observation

Froude had great powers of observation which he used to draw valid conclusions from very simple experiments, such as that with the tiny pendulum on a float, which led directly to his understanding of the dynamics of rolling. His measurements of the pressure drop along the length of the Torquay water main showed that accepted theories of friction between a rough surface and a liquid were wrong while measurements of friction on the launchways of the GREAT EASTERN cast doubt on the laws of friction between solid surfaces. This use of observation was not luck; as Louis Pasteur said 'In the field of observation, chance favours only the trained observer.'

15.4 Measurement

Observation gave Froude a starting point but the next step depended on accurate measurement. Froude's work as a judge of agricultural machinery (Chapter 7) shows his determination always to measure rather than to rely on subjective judgement. One may also see this passion for measurement in his letter to Newman of December 1859 (Page) in which he several times refers to the difficulty of measuring 'Faith'.

His work on agricultural machinery is an early demonstration of his ability as a clever designer of instruments and as a skilled mechanic with a great pride in his tools. These skills developed through his roll recorders to the complete instrumentation of the Torquay tank and, in particular, his propeller dynamometer. His instruments were very simple in concept; wood frames, brass wheels and leather driving bands but there was usually some fine detail to overcome some difficulty which he had appreciated, such as the need to get the line of action of a force in a specified direction. His experimental apparatus always exhibited excellently finished work where finished work was necessary and sufficiently though less finished in other parts. They worked and went on working, in some cases up till World War II.

He made many of these instruments himself as '... being a mechanic myself, and having a mechanic's shop, I have been able to save expense by preparing some of my own apparatus in a rough yet effective way, where, if a regular manufacturer had been applied to, it would have assumed an expensive character.' He pointed out that many great men, such as Sir William Thompson, were skilled in the making of inexpensive apparatus (ref 15.4).

15.5 Analysis

The measurements were then analysed with great care and examined for discrepancies. In the letter to Newman of 29th June 1853 Froude says, 'Science progresses only by following clues – it is only when, in such a pursuit, our stock of acknowledged principles fails to account for some unmistakeably residuary phenomenon, that experimentalists venture cautiously to think that they have really got hold of what may turn out to be a new and unacknowledged principle.' This determination to explain discrepancies is well shown both in his recognition that waves generated by the ship contributed to roll damping and by the care which he took to show that this explanation accounted precisely for the energy losses which could not be attributed to other forms of damping. Another example came in 1874 when he realised that the interaction between two ships close together could be important and, as a result, he re-examined the GREYHOUND trial results (Chapter 12).

Froude was clearly outstanding in mathematics, both at school and at university but his early work for I K Brunel can have given few opportunities to keep in practice. It is likely that his lost work on skew bridges had a mathematical basis and his development of the transition curve used some mathematics as did his paper on expansion but that was virtually all until about 1859. His 1861 paper on rolling was the most mathematical read to the new Institution of Naval Architects in its first decade. Froude credits Bell as an equal partner in this paper but, even so, Froude's contribution was a remarkable achievement for a man of fifty. It is usually accepted that mathematical ability peaks at an age of thirty or less and can only be maintained at a high level with constant use.

Froude described his mathematical background to the Royal Commission in 1872

> '...When I was at Oxford such mathematical teaching as had nominal reference to practical questions was infinitely too divorced from practice. I had, in a sense, to unlearn my Oxford teaching, when I began my profession as an engineer, and to start again on a sounder basis of thought; and, although, later in life, I found that the mathematics I had learnt were available, they had become very rusty.... I have found great advantage from it, and I ought not to be ungrateful for it,'

One may well feel that he was well taught for his learning to last so well.

In preparing Froude's obituary Henry Brunel challenged Froude's own view of his mathematical ability writing:

> 'Froude in public spoke slightingly of his mathematical knowledge and it wasn't of the highest order of the leaders of the present day but in private he spoke with gratitude of the way in which he'd been taught mathematics by his elder brother, Hurrell. His mathematics were tools that would work well, he understood their meaning and their use to make them do the most serviceable work.'

A real engineer's attitude to maths! Froude thought as an engineer, using mathematics, rather than seeing the maths as paramount, shown by his criticism of Bell for over reliance on formulae. His elegant, graphical analysis of rolling trials shows another aspect of his mind leading him to a usable solution for a problem otherwise intractable.

15.6 Simplification

The key to his later work on the powering of ships lies in his ability to break a complex problem down into simple components. Resistance was divided into frictional and residuary components whilst a little later the overall powering problem was split into resistance, propeller efficiency and interaction, each themselves sub-divided. It is clear, too, that Froude was well aware that these breakdowns were approximations which introduced errors but he satisfied himself that any such errors were either negligible or could be allowed for with satisfactory accuracy by a simple correction factor based on a comparison of model tests with trial results.

15.7 Trials

While Froude pointed out, forcefully and on numerous occasions, that full scale trials were too expensive to be used in the development of novel forms, he did see that they had a necessary part to play. In his evidence to the Royal Commission he said: 'Experiments on a small scale have satisfied me personally on this point, but in applying to a large scale the results of experiments on a small scale, there is always room for some doubt, and, at any rate public opinion is much more easily satisfied by experiments on a scale of 12 inches to the foot than on a smaller scale.'

His work with GREYHOUND showed that he recognised that the object of a trial must be set out clearly before starting and that a trial 'to see what happens' is almost always useless. GREYHOUND showed the beginning of what is now called 'ship-model correlation', a fancy name for the correction factor which is applied to model test results to bring them in line with trial data, an aspect which he and later Edmund developed from Denny's pioneer work. (Sometimes referred to as the 'fudge factor')

15.8 Partnership

Froude does not seem to have read widely on early work in his subjects: implying in his letter to I. K. Brunel of 16th August 1857 that he was unaware of any earlier work on rolling, but he kept closely in touch with contemporaries such as Rankine, Stokes and Denny. His friends made important contributions to his work such as that by Bell on the mathematics of rolling, Henry Brunel on many aspects of work whilst Mallock, Tower etc as well as his son Edmund all contributed. Others such as Pengelly, I. K. Brunel and, most important, Newman provided intellectual stimulus. Writings by all these friends make it quite clear that William Froude was the leader and, almost always, the source of inspiration.

15.9 William Froude – The Person

All William's friends emphasize his kindness. Henry Brunel refers to this in connection with the 'Polly' affair and, another time, wrote 'I should like to show you such a kind letter he wrote to me about a rude one I wrote him.' After his death, Henry wrote (1st June 1879)

'Since my father died nearly twenty years ago, Mr. Froude has shown not only the kindness of a father to me, but the confidence of a friend. I think I may say that I have his opinion or advice in every step of importance I have taken.'

This aspect also shows in the way in which he stood by Newman after the latter's conversion when most Anglican friends deserted him. It is clear that he treated all his contacts the same way, from workmen to Lords – Henry wrote:

'... such was his unfailing tact and consideration, that the desire of all was to assist and welcome him, indeed it was a pleasure to go an errand from him so cordial was the reception which mention of his name ensured.' (of Dockyard workers)

One may also see his ability to get on with people in an incident during his rolling trials on HMS DEVASTATION. On rejoining the fleet, the band of HMS SULTAN, on which he had previously carried out trials, struck up a music hall ditty of the day 'Willy, we've missed you.' Even today, scientists carrying out trials on warships may be seen as a bit of a nuisance and such feelings may have been stronger in the nineteenth century: William must have made a very favourable impression on the SULTAN.

The Devonshire Association obituary said of him that

'With such a character he brought brightness wherever he went' and continued '... one in any trouble or distress who met him, not knowing who or what he was, must have thought his life was spent in the tender concern for others which springs from forgetfulness of self, and a sense of the mystery of human life.'

This kindness to others may have contributed to his apparent indecisiveness, Henry writing 'Froude often thinks he's giving an order when his words only imply expressing an opinion'. His frequently expressed belief in 'The Sacred Duty to Doubt' may also have contributed to the appearance of indecision but this was only superficial, hiding an inner toughness as in overcoming virtually unanimous professional opposition to model testing.

He was a quiet, shy man; 'His voice had an almost pathetic tone, the outcome of a sympathetic heart'. His humour was laboured, almost schoolboyish. He often got mixed in speech, using the wrong words and the same problem occurs in his letters; even important ones, such as those to Reed on the tank proposal, have many crossings out. Letters were often wrongly dated making chronology difficult. One to Newman (5.8.64), which is even more illegible than usual, and somewhat incoherent in content, concludes with a postscript 'I have not time to read and correct this – if it is obscure, do not waste time in trying to make out its (meaning)' (I have guessed the last word). All this seems to show that he thought faster than he could communicate. His formal papers are very different, very clear, if a little long winded at times. His personal carelessness was part of acting too fast, he was always leaving things behind and apparently inclined to put his eyeglass in his mouth.

At university he had a reputation for practical jokes and there are occasional suggestions that this habit persisted. In October 1871, Henry Brunel was dozing peacefully on a couch at Chelston Cross and was awoken by his foot getting too warm finding that Froude was waving a red hot poker underneath it.

From his youth there are complaints of William's idleness, perhaps better seen as a lack of concentration on a single objective. Henry wrote that the Froudes can never find time for anything. At its best, this habit could be charming, 'Nothing was too small, nothing too great for his grasp. He would bring the same intense, yet almost playful, attention to the construction of a toy as to the analysis of the curves of an ironclad or the behaviour of an Atlantic wave.' But in completing the tank on time, it was irritating to his helpers. One may wonder how he achieved the respect of Isambard Brunel as the construction manager for the Exeter line.

On the other hand, Henry and other friends thought that he was overworking in the 1870s and pressed him to accept the offer of a voyage to South Africa as a well earned and much needed rest. Perhaps it was just that, at last, he had found something which grasped his whole attention.

Matters of religious belief were clearly a major concern and, increasingly, he saw faith as incompatible with scientific, questioning methods. His distress over Edmund's conversion and subsequent wish to become a priest shows that his views were very much stronger than those put forward in his more polite letters to Newman. But let his wife, Catherine, have the last word 'It is always extraordinary to me (seeing what excellent sense and judgement he has on most subjects) that in talking of Catholic matters, he does talk such nonsense.'

15.10 Marriage and Catherine

Catherine Henrietta Elizabeth Holdsworth Froude b 1809, d 24.7.78

> '...that amiable and gifted lady.'

There is very little material on her personality and almost all there is comes from the letters and journal of Henry Brunel. His friendship with Catherine endured till her death, punctuated by quite frequent minor quarrels. For example, in 1863, he wrote on 9th March

> 'Mrs. Froude was most cold and extraordinary in her treatment of me. I wonder what's up. I can't imagine.'

but on 16th March he was able to record

> 'Mrs. Froude was very cordial to me so I suppose I was mistaken as to her having intentionally snubbed me last week. If it was not

intentional she is a most extraordinary women.'

Henry was 20 at the time and there are indications that he was not always very tactful, quarrelling with William, too, at times. Two years later, Henry was to acknowledge her kindness as well as that of William during the 'Polly' scandal. He also wrote '...difficult to argue with Mrs. Froude because she doesn't believe in people, believes in her own power of judging things right and clear.' (7th June 1870)

Despite her strong intellectual interests, Catherine clearly had a lighter side. In 1877, Henry sent her the first weeks issue of the magazine 'Detective' which he said was better than Wilkie Collins and 'if a morbid taste likes it, it is to morbid taste.' Henry quite frequently met Catherine and Izy in London dining and visiting a show. There is another interesting comment 'Two Wards there; were nice and they undermined their sense of propriety adroitly and successfully with great assistance from the least expected – Mrs. Froude – her skill with the praiseworthy subject by means of subtle and intricate improper innuendoes defies description.' (22nd Jan 1869)

She was inclined to be forgetful and when in May 1871, Henry invited her to stay when next in London, he wrote to Izy by the same post warning her that her mother might put the letter in her pocket and forget it. Since William was also forgetful – and Henry worse – it must have been an interesting household.

Harper, presumably basing his remarks on Izy's memories, says that William's private life was 'a less than ordinarily happy one.' due to the religious differences between them. These differences were very strong and on a subject of great concern to both of them. During the mid 1850s William became increasingly concerned over the direction of Catherine's correspondence with Newman and persuaded Newman to stop writing though the correspondence was soon resumed. By 1854 William wrote to Newman:

> 'I fully believe that as far as reasonable or reasoning conviction goes, her judgement is against Catholicism – as far as feeling goes it is in favour – the feeling being partly what might be called fascination occasioned by the magnitude of the system, and what appears to her the adaptations of its ceremonial to her own peculiar turn of mind and partly her entire love and admiration for the Catholics she has known – a love and admiration which goes entirely beyond that which she feels for any other persons whatever.'

A study of Catherine's letters to Newman suggests that her view of Catholic faith had a very much stronger intellectual base than William realised – it would not be the last time that a husband failed to understand his wife. Her conversion and the practical difficulties of sharing a house with the Archdeacon of Totnes have been discussed in Chapter 3. There was an even greater emotional storm when Eddy first became a Catholic,

Chapter 15: William Froude – An Evaluation

influenced secretly by Catherine, and then wished to become a priest.

Extracts from two letters written much later when Newman wished to dedicate a book of his essays to William (6 August 1871) support this view.

> 'It is always extraordinary to me (seeing what excellent sense and judgement he has on most subjects) that in talking of Catholic matters, he does talk such nonsense.' and 'It seems to me that William is so utterly removed from the ordinary run of Sceptics; and his mistakes appear to me to proceed in great measure from crankiness, and a sort of over-scrupulousness.'

William's distress during her final illness and his enduring affection show in his letter to Newman of 23rd July 1878.

> 'Her entire unselfishness and affectionate thoughts for others, which have always been characteristic are, as I think you would expect, even more striking now...'

Her death affected him greatly and was to lead to his cruise to South Africa and thus his own death.

No portraits of Catherine have been located and there is no description of her – perhaps she took after her formidable father. Henry wrote in January 1875 'Never heard of Mrs. Froude being starved at school before – is that why she's so small?' Other than a few comments from Izy, her children have recorded no comments on their home life but all seem to have been successful and retained a happy relationship with both their parents.

15.11 Tributes

When William Froude died, there were many tributes paid to him. To conclude, a selection from these will be given, beginning with a slightly earlier one, on the award of the Royal Society's Royal Medal in 1876, the same year in which he was awarded the honorary degree of L.L.D. by Glasgow University.

> 'It is generally admitted that Mr. Froude has done more than anyone else towards the establishment of a reasonable theory of the oscillation of ships in wave-water, as well as for its experimental verification. ....He has also conducted a series of experiments, extending now over many years, on the Resistance, Propulsion and Form of ships, and on the very important and little-understood question of the law connecting the behaviour of ships, in all these respects, with that of models of ships on a much smaller scale.'

The Royal Society Obituary (drafted by Henry Brunel) said:

> 'Froude's character was one in which a rare degree of modesty and disregard of self was combined with a singular charm of voice and manner. This enabled him so to act in many difficult positions that he secured for himself a prompt recognition of the

great scientific knowledge and powers of mind with which he was endowed.'

Sir Henry Acland, M. D. wrote to the Times on his death:

'No workman in any art ever combined in juster proportion, few in so eminent a degree, the three properties of culture, science and practice.' (Other quotations from this letter have been used elsewhere.)

The Admiralty of the day was, as was usual, sparing in its praise but wrote to Edmund:

'I am commanded by My Lords Commissioners of the Admiralty to inform you that they have received, with a deep sense of sorrow, the intimation contained in your letter of the 26th instant of your father's death.

'They feel that Mr. Froude rendered great services to the Navy, and the country, in making his great abilities, knowledge, and powers of observation available for the improvement of the design of Ships, without reward or any other acknowledgement than the grateful thanks of successive Boards of Admiralty.

'My Lords desire to convey to you, and other members of the family, the expression of their most sincere sympathy at the irreparable loss which you have sustained a loss which cannot be looked on as other than a national one.'

References

15.1 W. Froude. On the Influence of Resistance upon the Rolling of Ships. Naval Science, London, 1874.

15.2 W. Froude. Report on the Exhibition of Implements at Falmouth. Bath and West of England Agricultural Journal, Vol. XVI, 1869.

15.3 Letter from J Spedding, April 1869. Contained in PRO file 116/137.

15.4 Report of the Royal Commission on Scientific Instruction and the Advancement of Science. Parliamentary Papers, 1874, XXII, p247-50.

Index

A

Abell v
Achilli 34
Acland, Dr. 215
Acland, Sir Henry 250
ACTIVE 170, 171
Actuator disk 226
Admiralty 55, 56, 179, 190, 204, 207
Admiralty Coefficient 113
Admiralty Experiment Works, Haslar, (AEW) i, iii, 67, 120, 192, 210, 216
Agricultural machinery 242
ALECTO 109
ANT class 189
Anthony 4, 12
Anti-rolling tanks 227
Appleton, E. 98
Appold, J. G. 95
Archdeacon 2, 11, 87
Archdeacon Froude 2, 15, 33, 93
Archdeacon of Totnes 1
ARCHIMEDES 108
Armathwaite Hall 15
Armstrong, Sir W. G. & Co 89
Armstrong, William 90
Arthur 27, 83
Attwood 52
AUDACIOUS 45

B

Badden 20
Baker 14
Baker, G. S. 233
Barnaby 40, 57, 66, 127, 140, 202
Barnaby, Nathaniel ii, 65, 174
Barnes 56, 128
Bassenthwaite 31
Baxter, W. E. 134
Beaufoy 106, 112, 117, 125, 178, 191
Bell 37, 40, 46, 47, 51, 245
Bell, William 39
Bidder 29, 93, 127, 171
Bidder, G. P. 139

bilge keels 48, 59
BOADICEA 209
Bosanko, Mr J. v
Bourguer 105
Brass model 223
Brenden, Piers 11
British Association 95, 120, 135, 163, 178, 183, 240
British Association Committees 110
Brompton Oratory 220
Brown, John 232
Brunel 18, 20, 32, 37, 39, 80, 163
Brunel family 87
Brunel, Henry iii, 14, 23, 29, 37, 43, 49, 55, 60, 67, 76, 79, 83, 88, 98, 109, 117, 127, 135, 146, 158, 171, 190, 208, 215, 244
Brunel, Isambard Kingdom iii, 25, 46, 87, 109, 221, 245
Brunel, Mary 88, 100
Brunels 90

C

Calibration Runs 153
CAPTAIN 53, 55, 130
Catherine 17, 33, 71, 93, 209, 248
Catherine's Conversion 35
Centenary Open Days i
Challenge 5
Channel Tunnel 95
Chappell 108
Chelston Cross 80, 87, 127, 205, 215
Chief Constructor of the Navy 128, 164
Childers 57, 129
Chiswick 220
Claxton 109
Clydebank 232
Cockington 97
Coles, Cowper 55
COMET class 189
Commission on the Advancement of Science 163
Committee 59
Committee of Design of Ships of War 165

Committee on Designs 57, 147
CONQUEST 198
Constructor's Department 167
Controller 138
Controller, Sir Spencer Robinson 56
Cowley Bridge 17
Crimean War 221
CROCODILE 61
Crossland 47, 127
Cullompton 19
CYCLOPS 58

D

d'Alembert 105
Dartington 3, 33, 40, 87, 210
de Lome, Dupuy 112
Denny, William 177, 232
DEVASTATION 58
Devonshire Association v
d'Eyncourt 229
Dirkzwager 149
DNC 199
Donkin 78
DREADNOUGHT 231
du Monceau, Duhamel 105
Duke of Somerset 56
DUKE OF WELLINGTON 46
DWARF 109
Dynamometer 149, 170
Dysentery 210

E

E in C 199
Eddy 76, 94, 201
Eddy Making Resistance 184
EDINBURGH 217
Edmund ii, 27, 55, 71, 90, 122, 132, 148, 171, 201, 215
Eels 224
Eliza 27, 71, 87
Eliza (Izy) v
Eliza Margaret 78
Elmsleigh 88
Engine and Governor 151
Ericsson 108
Exeter 20

F

Fairbairn 126
Fall of ends 189
Farm machinery 28
Finance 161
Fincham 106
First submarine model 230
Flying machine 93
Flying staircase 99
FRANCIS SMITH 108
Froude 17, 30, 40, 55, 71, 89, 106, 117, 127, 145, 170, 177
Froude (ill) 20
Froude (railway career) 20
Froude, Anthony 201
Froude, Catherine 72
Froude, Catherine Henrietta Elizabeth Holdsworth 247
Froude dynamometers 28
Froude, Edmund vii, 67, 89, 114, 138, 146, 195, 231
Froude, Hurrell v, 75, 90
Froude, Margaret 14
Froude, Mary v
Froude, Mr ii
Froude, Mrs 79
Froude, Mrs William 8
Froude, Mrs. 248
Froude Number 239
Froude, Phillis 15
Froude, R. E. 63, 228
Froude, Robert 3
Froude, William ii, 19, 71, 87, 111, 122, 163, 182, 232, 239
Froude's Law 121
Froude's Law of Comparison 48, 239
Froude's model tank 67
Froudes, William 19

G

Gawn, R. W. L. viii, 231
Gem 94
GLATTON 58
Gooch, Sir Daniel 148
Goodall, Sir Stanley vii, 224
Goulaev 201
Governor 152
Graduation 9

Graph paper ruling machine 158
Gravatt 18
Gravatt, William 17
GREAT EASTERN 36, 39, 88, 113, 127, 217
Great Exhibition 87
GREYHOUND 60, 112, 148, 169, 177, 239
Guiney, L. E. v
Gunboat Yard 221

H

Harper v, 71
Haslar viii, 182, 220
HB 63
Henry vi, 60, 79, 89, 136, 165, 201
Henwood, C. F. 140
Henwood, Edwin 140
Holdsworth, Arthur Howe 15, 27, 78, 93
Holdsworth, Catherine Henrietta Elizabeth
    iv, 27
Holdsworth Family 15
Hull Efficiency Elements 195
Hull Form 185
Hurrell 3, 27, 71, 87, 160, 201, 215
Hurrell (elder brother) 2
Hurrell (letters) 5
Hurrell, Richard 11
Hurrell, Robert 11

I

Inclining Experiment 51
INFLEXIBLE 67, 162, 186, 227
Institution 48
Institution of Mechanical Engineers 101
Institution of Naval Architects (INA) 42, 128,
    171, 183, 215
IRIS 162, 198, 223
Iron cross 210
Isambard iii, 27, 87, 109
Isambard (jnr) 209
Iso-(K) books 219
Izy 27, 92, 201

J

James Anthony 2

K

Keble 8
King's Scholarship 5

Kittoe, G. D. 24, 92, 149, 170
Krylov Institute 232

L

Laird 56
Lake Bassenthwaite 4, 107, 117, 190
Lanchester, F. W. 226
Launching problems 36
Law of Comparison 117
Lerbs 192
Littlemore 33
Lloyd, Thomas 32, 108
Lowndes 5

M

Maby, Mr G. E. vi
Maby, Mr R. G. vi
Mallock Family 14
Mallock, Henry Reginald Arnulph v, 161,
    226, 245
Mallock vibration recorder 226
Mallock, William 5, 97, 145, 220
Margaret 4
Margary, P. J. 148
Mary 8, 27, 55, 87, 209
Mary Catherine 80
Maudslay 198
MEDEA 227
MEDUSA 14
MERKARA 196
Merrifield, C. W. 111, 124, 127, 170
MERRIMAC 55
Metacentre 50
Metford 14, 92, 209
Methodical Series 218
Milne, Admiral 65
MONARCH 56
MONITOR 55
Moriarty, H. A. 51
Mosely, Canon 47, 137
Mozley 7
Mumford 232

N

National Physical Laboratory 232
Naval Science 128
Newman, John Henry (Cardinal) iii, 21, 33,
    220, 241

Newton, Isaac 73, 105, 120, 140
Newton's theory of gravity 73
Nobles 90
Nominal Horse Power 108

O

Oriel 7

P

Paddington 20
Paignton 46, 87
Pakington 56
Palmer iii, 17, 106
Particulars of the Models 120
Patent 24
Paul, H. v
Payne, M. P. viii, 230
Pengelly 102
Perret, J. R. 159, 232
PERSEUS 60
Phillip 64
Phillis 8
"Plank" Tests 178
POLYPHEMUS 188
Popoff, Admiral 187
Propeller 177
Propeller Design 225
Propeller dynamometer 193
Propeller Performance 190
Purvis, F. P. 159, 232

R

RAMUS 206
Ramus 124, 154
RAMUS Form 204
Ramus, Reverend C. M. 204
Rankine 31, 112, 128, 174, 190, 245
Rankine, Professor 166, 178
RATTLER 32, 108
RAVEN 94, 118
Rawson, Robert 106, 140
Reech 111, 125
Reed, Sir Edward iii, 31, 45, 55, 101, 128, 174, 177
Rendel 67
Resistance of flat planks 179
Rewe 17
Righting lever 53

Righting moment 50
Robert 3
Robert Edmund 78
Robert Hurrell 1
Robinson, Sir Spencer 56, 135
Rocket propelled ram 208
Roll recorder 62
Rolling 39, 55, 227
Rorison 92
Rota 232
Royal Commission on Scientific Education 68
Royal Society, Fellowship of the 102
Royal Society Obituary, The 249
ROYAL SOVEREIGN 68
Russell, J. Scott 37, 47, 55, 106, 117, 139, 178, 217

S

Science Museum, Wroughton 146
SERAPIS 61
SHAH 67, 162, 200
Ship Dynamometer 197
Ship Tank 144
Simon's Town 210
Skin Friction Correction 182
Smith, Petit 108
SNAKE 189
Soaring of Birds 208
Society for the Improvement of Naval Architecture 106
Somerset, Duke of 128
South Devon Atmospheric Railway 24
Spedding Family 15
Spedding, H 209
Spedding, James 31, 132
Spedding, John 4
Spedding, Margaret iv
Spedding, Phillis 12
Spedding, Thomas Story 15
Speed trials 177
Spezia Tank 232
St Albans 232
St Petersburg tank 232
Stability 50
Stability at Large Angles of Heel 52
Standard Models 223
Stanhope, Earl of 106
Stearine 131

Stokes, Professor 178
SULTAN 64
SWAN 94, 118, 129
SWAN & RAVEN 55, 87, 117, 127, 177, 239

T

Tank at Torquay 66, 143, 179
Taylor, David W. 220
Thames Iron Works 89
The building 145
The Law of Comparison 182
Theory of propeller action 225
Thompson, Sir William 166, 178
Tidemann, Dr. B. J. 149, 233, 240
Torquay viii, 55, 160, 182, 216
Torquay water main 125, 178
Tower, Beauchamp 14, 29, 93, 146, 208, 222
Tracey, Mr G. vi
Tractarian Movement 11
Trials 196
Tributes 249
Trustee for several turnpikes 27
TRUSTY 55
Tudor, G. D. C. 87

V

Vickers 232
VIRGINIA 55
von Hugel, Baron Anatole 216
von Hugel, Baroness v
VULCAN 220

W

Wake 31
WARRIOR 88, 108, 128, 192, 217
Watcombe 87
Watts, Isaac 128
Watts, Phillip 60, 159, 171, 227
Wave maker 228
Wave Making Resistance 182
Wave pattern generated 183
Waveline Theory 113
Wax Model Making 155
Wax models 157
West Somerset and Weymouth 20
Westmacott, Percy 90
Westminster School 5
White, Sir William ii, 68, 219, 232

William 5, 76, 87, 159, 201, 215
William Froude 188
William's eyes 21
Wooley 174
Wooley, Dr. 39, 67, 135, 166
Wooley, Rev. Joseph 47
Wright, Dr. Tom 106

Y

Yacht Racing Rules 233
Yachting 93
Yarrow, Sir Alfred 232